Agriculture in Developing Countries

Due to intrinsic and location-climate-specific advantages, most farmers in the tropics quite often base their decisions more on their own knowledge. However, this does not in any way, undermine the contribution of the scientists to agricultural technology, which farmers do need. The implication is that the farmers' indigenous knowledge and modern scientific knowledge are not substitutes, but are complimentary towards achieving the objective of improving the welfare conditions of the farming community. It is in this context that this edited volume impresses upon us that there is an urgent need for new approaches to agricultural technological development to be built upon indigenous knowledge, so that technology can strengthen farmers' ability to innovate and to provide useful information for further development of technology. Such participatory development of farmers in tailoring new technologies by scientists needs to be strengthened in the new era of agricultural technological development to combat poverty around the world. The manner in which farmers in Asia have been making successful efforts to adjust to the constantly changing agricultural environments towards realizing their potential outputs, using appropriate farm level primary data, is the subject matter and the central idea of this book.

Agriculture in Developing Countries

Technology Issues

Edited by

Keijiro Otsuka
Kaliappa Kalirajan

⊗SAGE Los Angeles • London • New Delhi • Singapore
www.sagepublications.com

First published in 2008 by

 SAGE Publications India Pvt Ltd
B-1/I-1, Mohan Cooperative Industrial Area
Mathura Road, New Delhi 110 044, India

SAGE Publications Inc
2455 Teller Road
Thousand Oaks, California 91320, USA

SAGE Publications Ltd
1 Oliver's Yard
55 City Road
London EC1Y 1SP, United Kingdom

SAGE Publications Asia-Pacific Pte Ltd
33 Pekin Street
#02-01 Far East Square
Singapore 048763

Published by Vivek Mehra for SAGE Publications India Pvt Ltd, typeset in 10.5/13 pt Aldine401BT by Star Compugraphics Private Limited, Delhi and printed at Chaman Enterprises, New Delhi.

Library of Congress Cataloging-in-Publication Data
   Agriculture in developing countries: technology issues/edited by Keijiro Otsuka and Kalippa Kalirajan
      p. cm.
   Includes bibliographical references and index.
      1. Agriculture innovations—Developing countries.   2. Sustainable agriculture—Developing countries.   3. Farmers—Developing countries— Attitudes.   4. Agriculture—Research—Developing countries.   I. Otsuka, Keijiro.   II. Kalirajan, K.P.

S494.5.I5A3263	630.9172'4—dc22	2008	2008012179

ISBN: 978-0-7619-3662-6 (HB)   978-81-7829-828-3 (India-HB)

The SAGE Team: Sugata Ghosh, Parikshit Bhardwaj, Amrita Saha and Trinankur Banerjee

Contents

List of Tables   7
List of Appendix Tables   10
List of Figures   11
Preface   13
Acknowledgements   15

1. Technology Issues in Developing Countries' Agriculture   17
    Keijiro Otsuka and *Kaliappa Kalirajan*

2. Economic Assessment of an Indigenous
    Technology in Bangladesh   28
    Md Abdus Samad Azad and *Mahabub Hossain*

3. Research Investment on Technology Development
    in Peninsular India   46
    G.S. Ananth, P.G. Chengappa and *Aldas Janaiah*

4. Impact of Forage Options for Beef Production
    on Small Farms in China   61
    N.D. Macleod, S. Wen and *M. Hu*

5. Optimal Land-use with Carbon Payments
    and Fertilizer Subsidies in Indonesia   81
    Russell M. Wise and *Oscar J. Cacho*

6. Deterioration of Tank Irrigation Systems
    and Poverty in India   100
    Kei Kajisa

7. Agricultural Technology and Children's Occupational
    Choice in the Philippines   113
    Keijiro Otsuka, Jonna P. Estudillo and *Yasuyuki Sawada*

8. A Panel Data Model for the Assessment
of Farmer Field Schools in Thailand 137
Suwanna Praneetvatakul and
Hermann Waibel

About the Editors and Contributors 152
Index 158

List of Tables

2.1 Socioeconomic Characteristics of the Sample Farmers 37

2.2 Variation in Agronomic Parameters Practised in
Rice Cultivation under *Bolon* and *Naicha* Systems 39

2.3 Economics of Aman Rice Cultivation under
Bolon and *Naicha* Systems at Northern Districts
of Bangladesh 40

2.4 Difference in Yield and Other Production Parameters
between *Bolon* and *Naicha* Systems 40

2.5 Maximum Likelihood Estimates of the Stochastic
Frontier Production Function 41

2.6 Determinants of Technical Inefficiency Model 42

2.7 Technical Efficiency Estimates of the Rice Farms
under *Bolon* and *Naicha* Systems 42

3.1 Land Saving Effect Due to Technology in Major Field
Crops of Karnataka (1970–71 to 1994–95) 53

4.1 Selected Data from a Census Survey of Households in
Chunquitang Village, Dao County, Hunan Province
for 2002 68

4.2 Assumed Characteristics of Synthetic Household,
Chungquitang Village, Dao County, Hunan Province,
Used for Scenario Modelling 68

4.3 Feed Year Plan for Providing Forage for Small-holder
Cattle Production (3 Head) in Hunan Province 69

4.4 Selected Household Parameter Values
for Scenario Modelling 71

4.5 Revenue, Costs and Profit Measures for Scenario 1 73

4.6 Revenue, Costs and Profit Measures for Scenario 2 74

4.7 Revenue, Costs and Profit Measures for Scenario 3 75

4.8 Revenue, Costs and Profit Measures for Scenario 4 77

5.1   Base-case Parameter Values   87
5.2   Base-case Values (Coefficients) for the Dependent
       Variables of the Quadratic Equations Defining the
       Biophysical Numerical Model   88
5.3   Four Base-case Carbon- and Fertilizer-price Scenarios
       Simulated in the Dynamic-programming Model: Each
       of these is Simulated for a 15 per cent Discount Rate   89
5.4   Optimal Decisions Over Five Cycles for Base-case
       Fertilizer- and Carbon-price Scenarios, and Two High
       Fertilizer-price Scenarios (Scenarios 5 and 6), at a Low
       and High Initial Soil–Carbon Level, and a Discount
       Rate of 15 per cent   94

6.1   Comparison of Rice Yield, Income, Consumption
       Value and Subjective Poverty Assessment of Sample
       Households between Collective Management Inactive
       Villages and Active Villages   104
6.2   Comparison of Rice Yield between Different
       Irrigation Statuses   106
6.3   Comparison of Income Per Capita between
       Different Irrigation Statuses   107
6.4   Comparison of Consumption Expenditure Per Capita
       between Different Irrigation Statuses   108
6.5   Comparison of Profit on Rice between Different
       Irrigation Statuses   109

7.1   Characteristics of Sample Households in the Study
       Villages in the Philippines, 1985/89 and 2001/04   119
7.2   Characteristics of the Children and Parents in the Study
       Villages in the Philippines, 1985/89 and 2001/04   120
7.3   The Determinants of Completed Years in School of
       Children in the Study Villages in the Philippines, 1989
       and 2004 (Ordinary Least Squares)   123
7.4   Determinants of the Probability of Children to
       Engage in Non-farm Work in the Study Villages
       in the Philippines, 1989 and 2004 (Probit)   123
7.5   The Determinants of Farm and Non-farm Income
       in the Study Villages in the Philippines,
       1985 and 2004 (Ordinary Least Squares)   126

7.6 The Determinants of the Incremental Years in
School of Children in the Study Villages in the
Philippines, 1985 and 2004   130

7.7 Determinants of Credit Constraints in the Study
Villages in the Philippines, 2001   132

8.1 Summary of Short-term Linear Shift
Effects from FFS Training   144

8.2 Impact of FFS on Pesticide Expenditures and Environmental Impact Quotient in the Short-term, Two Periods
Growth Model   145

8.3 Impact of FFS on Pesticide Expenditures and Environmental Impact Quotient in the Long-term, Three Periods
Panel Data Growth Model   145

List of Appendix Tables

3.1 Total Factor Productivity Growth of Major Crops
in Karnataka State   56

3.2 Number of Varieties Developed and Released in
Major Crops by SAUs in Karnataka (1970–71 to
1994–95)   58

3.3 Number of Technologies, Other than Varieties
Developed and Released in Major Field Crops by
SAUs in Karnataka (1970–71 to 1994–95)   58

3.4 Discipline-wise Number of Technologies, Other than
Crop Varieties Developed and Released in Major
Field Crops by the SAUs of Karnataka
(1970–71 to 1994–95)   59

8.1 Socioeconomic Characteristics of Sampling
Farmers by Group, before the FFS
Training, Cropping Year 1999/2000   147

8.2 The Average Differences in Selected
Performance Indicators for Short-term Impact   149

8.3 Scenario Testing for the Farmer Net
Benefit for the Short-term Impact   150

List of Figures

1.1 Sources of Output Growth   21

2.1 Aman Rice Cultivation under the *Bolon* System   29
2.2 Daily Average Rainfall Data, 1992–2003 in the
Study Areas   30

4.1 Soil Map of South China (after Xi et al., 1990)   62
4.2 Structure of the Economic Model   65

5.1 Pareto Efficient Production Possibilities of an Individual
Landholder when (1) not Receiving Payments for Positive
Environmental Externalities (E_1) and (E_2) when Positive
External Effects are Internalized through Carbon-
sequestration Payments (E_2) 82
5.2 Optimal Management Regimes Obtained by Solving
the Dynamic-programming Model for Four
Combinations of Fertilizer and Carbon Prices, at
Base-case Parameter Values   90
5.3 Optimal State Paths Associated with the Optimal
Management Decisions Obtained by Solving the
Dynamic-programming Model for Four Combinations
of Fertilizer and Carbon Prices and Two Levels of
Initial Soil Carbon, at Base-case Parameter Values   92
5.4 The Trajectory of the Eligible–Carbon Stock Associated
with the Optimal Management Regimes for the Different
Prices of Carbon and Fertilizer for a Poor Quality Soil   93

7.1 Hypothesized Interrelationships among Modern
Agricultural Technology, Schooling Investments
and Non-farm Employment   116

Preface

Recently, our colleague, Yujiro Hayami, in his Presidential Address to the Fifth Conference of the Asian Society of Agricultural Economists held in August 2005 in Iran argued that governments in middle income countries are confronted with two competing issues of securing low-cost food to urban consumers and supporting the income of farmers in rural areas, which force them to tinker with agricultural policies widening the income gap between farm and non-farm workforce. He further called for more research in this area to prevent such a widening of income gap between farm and non-farm population, which has the potential to lead to several unwanted socio-economic problems derailing the growth momentum. It is in this context that this edited volume examines what new directions need to be emphasized in designing agricultural policies to accelerate and maintain sustainable agriculture production and agricultural income growth. Such examinations would contribute to our understanding of methods to reduce income disparities not only within agriculture, but also between rural and urban areas.

Technology plays a central role in agricultural policies. Demands for agricultural technology are now changing and diversifying in the face of rapid economic transformation in Asia. New approaches, for instance, involving the combination of frontier sciences with the practical indigenous knowledge of farmers, such as new crop-tree and crop-cattle technologies as well as production systems approaches to breaking through yield plateaus, are needed. It becomes imperative for the application of scientific research, particularly for unfavourable areas, where most poor people live, while taking into account the conventional wisdom in the form of indigenous knowledge of farmers. And in all areas, as agricultural demands have lead to the intensified resource use, greater attention is needed to sustainability and the containment of potentially adverse environmental impacts.

Drawing on the above arguments for the need for exploring agricultural technologies and better management practices at the farm level to improve and sustain agricultural growth, and thereby reduction in poverty in Asia, we have compiled a set of seven quality research papers that were submitted for presentation at the 26th Conference of the International Association of Agricultural Economists (IAAE) held in August 2006 in Gold Coast along with an overall introduction to this special volume. Based on the quality of their papers, these authors received financial support to attend the conference directly from the National Graduate Institute for Policy Studies in Tokyo. We do hope that this edited volume contributes to the achievement of the goal of the Conference, 'Contribution of Agricultural Economics to Emerging Policy Issues', and that the contents of this book will be useful for the farming community, researchers and policy makers.

Keijiro Otsuka
Kaliappa Kalirajan

Acknowledgements

We acknowledge and appreciate the financial support given by the National Graduate Institute for Policy Studies (GRIPS) in Tokyo under the 21st Century Centre of Excellence Program to the authors who attended the 26th Conference of the International Association of Agricultural Economists (IAAE) held in Gold Coast in . August 2006, and to the publication of this edited volume by SAGE Publications in New Delhi. We sincerely thank the GRIPS administration, particularly Ms. M. Mukai, Associate Professor T. Yamano, and the participants in our sessions at the IAAE conference for their excellent support given to us at various stages of the production of this book. Editorial assistance provided by Ms. M. Tanaka is appreciated with thanks. We thank SAGE Publications for their efforts in bringing out this book elegantly to the readers.

Acknowledgements

W

1

Technology Issues in Developing Countries' Agriculture

KEIJIRO OTSUKA AND KALIAPPA KALIRAJAN

Introduction

More than two-thirds of the national workforce even today depends, directly or indirectly, on agriculture which continues to generate about one fourth of the GDP in several developing countries. Increased agricultural productivity and profitability, especially in resource-poor areas, are crucial to rural poverty alleviation and improved welfare of rural women and other disadvantaged groups. It is in this context that exploring the possibilities of new technology and better management practices at the farm level play a vital role in both national and international agricultural research. Generally, agricultural technological innovations have been a 'top-down' approach in the sense that scientists design the experiments concerning a new technology in their laboratories and on farmers' trial fields. Though scientists consult farmers at some stages while designing trials, mainly laboratory results dominate the prescription of the technology to farmers. In the early stages, in order to motivate the farmers to adopt the recommended practices of the new technology, national governments provided subsidies to purchase new inputs such as fertilizers and seeds which made the prices of fertilizer and seeds cheaper relative to rice prices and other cereal prices

(Hayami and Ruttan, 1985). Though such incentives have yielded substantive results, particularly in the context of 'Green Revolution' technology transfer, in terms of high levels of adoption within short periods, literature has shown that there have been significant differences between farmers' yields and scientists' experiments at the field level (IRRI, 1979, 2006).

The existence of such 'yield gaps' at the field level consistently in several countries raised concerns among plant breeders, social scientists and policy makers. As a consequence, different 'constraints projects' were carried out during the late 1970s and 1980s and periodically thereafter by both national and international organizations. These studies concluded that in most cases farmers were not following the recommended techniques and amount of inputs and that several social, non-price and organizational factors constrained farmers from using the recommended practices of the chosen technology (IRRI, 1979). Such conclusions highlighted the fact that the 'wholesale' transfers of agricultural technology from the technology shelf to farmers' fields would not yield the expected experimental stations' results. National agricultural research institutes started working on 'location specific' varieties using the modern varieties developed at international research organizations such as the International Rice Research Institute (IRRI) as parental materials and drawing on farmers' knowledge about their location specific characteristics (Otsuka and Kalirajan, 2006). These location specific varieties have been characterized by higher-grain quality and wider geographical adaptability because of the development of a large number of location-specific varieties (Janaiah et al., 2006).

Nevertheless, the location-specific research also followed more or less the 'top-down' approach and still 'yield gaps' at field levels persisted. In India, yield growth in Green Revolution crops—rice and wheat—started to slow down in intensively irrigated ecosystems in the 1990s (Bhalla and Hazell, 1998). Further, yield gaps of a different kind, which may be named as 'farmers' own yield gaps', defined as the difference between what is perceived as farmers' potential and farmers' actually realized outputs within a homogenous group of farmers, attracted the attention of researchers and policy makers (Kalirajan, 1990). Though the national and international research institutions paid more attention to closing the gap between experimental stations' yields and farmers' field-level yields, much less attention has been given towards closing 'farmers' own yield gaps'. The existence of such 'farmers' own yield gaps' at farmers' field levels indicate the need for a change in the strategy for the improvement of agricultural technology because farmers now require technologies which are sustainable under their production environments without causing much negative impact on the ecological environment. New approaches, for example, those combining frontier sciences with indigenous crop-tree and crop-cattle technologies as well as production-systems approaches to breaking through yield plateaus— are needed. It becomes imperative for the application of scientific research for particularly unfavourable areas while taking into account the conventional wisdom of farmers. And in all areas, as agricultural demands on the resource base intensify, greater attention is needed for sustainability and the containment of potentially adverse environmental impacts. It is now argued in the literature that the effective supply to meet the demands for new agricultural sustainable technologies may not be possible solely from scientific knowledge, but from a combination of scientific and 'indigenous knowledge' of the farming community.

Drawing on the above arguments about the need for exploring agricultural technologies and better management practices at the farm level, while taking into account farmers' experiences to improve and sustain agricultural growth in Asia, we have compiled a set of seven quality research papers that were submitted for presentation at the 26th International Association of Agricultural Economists (IAAE) conference in Gold Coast in Australia from 12–18 August 2006, under the sponsorship of Centre of Excellence (COE) project of the National Graduate Institute for Policy Studies. In addition, this set of papers provides a policy-setting framework for examining the technology issues that are relevant in the present context of developing countries' agriculture. The following section discusses the theoretical link between output growth and technology, and how 'farmers' own yield gaps' emanate from their use of a chosen technology. The next section discusses the need for improvement in the approach to developing new technologies by combining both scientific and indigenous knowledge. How farmers in certain developing Asian countries have adapted the existing new technologies through their farming experiences to best suit their production environment are briefly described in the form of summary of papers that are included in this special volume.

Output Growth, Technology and 'Farmers' Own Yield Gaps'

Evaluation of agricultural growth performance can be made in several ways. One important method is to identify the sources of output growth so as to gauge whether it is the increase in inputs or increase in input-use efficiency, or improvements in technology that contributed more to output growth. In accounting for output growth, the conventional Solow 'residual' approach fails to recognize and estimate effectively the key role of technological change within the components of total factor productivity (TFP) growth. As technological change is obtained as the 'residual', it is a 'catch all' measure not only for technological progress but also for other factors such as missing inputs and quality variations in inputs. Further, at any point in time, total factor productivity is the combined result of technological progress and technical efficiency, or the efficiency with which factors are used, given the technological progress, to produce outputs (Kalirajan, 1990). From the perspective of long-run agricultural policy, it is crucial to distinguish the increment in productivity that occurs from technological progress from that which results from improved technical efficiency in the application of already established technologies.

Drawing on Kalirajan et al. (1996), Figure 1.1 illustrates the decomposition of total output growth into input growth, technical progress and technical efficiency improvement. In periods 1 and 2, the farm-firm faces production frontiers F_1 and F_2 respectively, which are feasible under the farm-firm's production environment and are different from experimental stations' frontiers. If a given firm has been technically efficient under its own production environment and operates on its frontier, the output would be y_1^* in period 1 and y_2^* in period 2. On the other hand, if the firm is technically inefficient and does not operate on its frontier due to various non-price and organizational factors, such as tenure insecurity (Hayami and Otsuka, 1993; Otsuka and Place, 2001), then the firm's realized output is y_1 in period 1 and y_2 in period 2.

Technical inefficiency (TE) is measured by the vertical distance between the frontier or potential output that is feasible under the given technology and the realized output of a given firm, that is, TE1 in period 1 and TE2 in period 2, respectively. These TE1 and TE2 are 'farmers' own yield gaps' in period 1 and period 2, respectively.

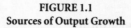

FIGURE 1.1
Sources of Output Growth

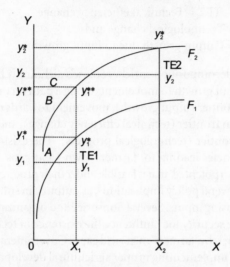

Hence, the change in technical efficiency over time is the difference between TE1 and TE2. Now, suppose there is technological progress due to scientific innovations carried out at national and international research institutes, then a firm's potential frontier with respect to its own production environment shifts to F_2 in period 2. If the given firm keeps up with the technical progress, more output is produced from the same level of input. So the firm's output will be y_1^{**} from x_1 input shown in the figure. Technological progress in this paper is measured by the distance between two frontiers $(y_1^{**} - y_1^*)$ evaluated at x_1. Denoting the contribution of input growth to output growth (between periods 1 and 2) as Δy_x, the total output growth, $(y_2 - y_1)$, can be decomposed into three components: input growth, technological progress and technical efficiency change.

Referring to Figure 1.1, the decomposition can be shown as follows:

$$
\begin{aligned}
D &= y_2 - y_1 \\
&= A + B + C \\
&= [y_1^* - y_1] + [y_1^{**} - y_1^*] + [y_2 - y_1^{**}] \\
&= [y_1^* - y_1] + [y_1^{**} - y_1^*] + [y_2^* - y_1^{**}] - [y_2^* - y_2] \\
&= \{[y_1^* - y_1] - [y_2^* - y_2]\} + [y_1^{**} - y_1^*] + [y_2^* - y_1^*] \\
&= \{TE1 - TE2\} + TC + \Delta y_x
\end{aligned}
$$

where

$y_2 - y_1$ = Output growth
TE1 − TE2 = Technical efficiency change
TC = Technological change and
Δy_x = Output growth due to input growth.

The above decomposition enriches Solow's dichotomy by attributing observed output growth to movements along a path on or beneath the production frontier (input growth), movement towards or away from the production frontier (technical efficiency change), and shifts in the production frontier (technological progress). The existence of technical inefficiencies leading to 'farmers' own yield gaps' with respect to 'farmers' own potential' that is feasible under their production environments bears several policy implications, as output can still be increased without increasing inputs. Several non-price and organizational factors such as tenure security may influence the existence of technical inefficiencies and thereby 'farmers' own yield gaps'. Government intervention in the form of implementing proper agricultural development policies may nullify the negative effects of these non-price and organizational factors on 'farmers' own yield gaps'. Nevertheless, agricultural research has the potential to significantly reduce the negative effects of 'farmers' own yield gaps' on their income level by providing them with more opportunities in the form of combining crop-tree and crop-cattle technologies. It is in this context that combining farmers' indigenous knowledge with the scientific knowledge of scientists to arrive at a number of feasible technologies to improve farmers' income level becomes crucial.

The Need for Combining Indigenous and Scientists' Knowledge

The environment of agricultural production has been changing constantly due to various factors such as variations in governments' macroeconomic policies and climate changes. This means that there is a need for experimental approach of constantly testing new possibilities and combinations to match the changing environment. It may be difficult for scientists with a formal approach to technology development to keep up with the constantly changing production environment in several countries and to supply suitable agricultural technologies towards sustaining farmers' income levels. It is in this context that the farmers' indigenous knowledge becomes critical in tackling some of the issues that may arise due to changing agricultural production environment. For example, as a consequence of the constantly changing production environment due to climate changes in the Mymensingh region of Bangladesh, farmers have been constantly adjusting the sequence of cropping, planting sometimes three, sometimes two, crops a year, and incorporating new varieties as they become available. Scientists can take note of this indigenous knowledge while developing crop varieties to suit such variations of production environment elsewhere. When the technology consists of more complex farming systems, understanding indigenous knowledge of the farmers is important. For example, agroforestry and other intercropping farming systems are so complex that the formal experimental approaches of scientists have proven to be less realistic. On the other hand, farmers, who have the potential for evaluating these kinds of systems more accurately, understand and adapt their farming strategies to meet changing needs.

Due to these intrinsic and location-climate-specific advantages, most farmers in the tropics quite often base their decisions more on their own knowledge. However, this does not, in any way, undermine the contribution of the scientists to agricultural technology, which the farmers do need. The implication is that farmers' indigenous knowledge and modern scientific knowledge are not substitutes for each other but are complimentary towards achieving the objective of improving the welfare conditions of the farming community. Particularly, in order to minimise the negative impact of 'farmers' own yield gaps' on their income levels—even though changes may be needed in macroeconomic polices—there is urgent need for new approaches to agricultural technological development in the form of providing more opportunities than a single-crop-production technology. Such new approaches to agricultural technology need to be built upon indigenous knowledge, so that the technology strengthens farmers' ability to innovate and to provide useful information for further development of technologies. Such participatory development of farmers in tailoring new technologies by scientists needs to be strengthened in the new era of agricultural technological development to combat poverty around the world.

Farmers' Indigenous Knowledge and Technology in Asia

Unlike technological changes in manufacturing industries, realization of potential outputs from agricultural technologies depends largely on farmers' adjustment efforts in production, compounded by the complexity and uncertainty associated with the climate changes. Discussed in the following pages is an exposition of the successful efforts of farmers in Asia to adjust to the constantly changing agricultural environment towards realizing their full potential output. A brief summary of the papers is given below.

In Chapter 2; Abdus Azad and Mahabub Hossain, using a case study from Bangladesh, have demonstrated how important it is to develop scientific research for unfavourable areas while taking into account the conventional wisdom of farmers. The authors have assessed the profitability and economic efficiency of the 'double transplantation' system of crop establishment, which is an indigenous technology developed by farmers to manage the risk of submergence in flooded areas, compared to the single transplantation system commonly practiced elsewhere in Bangladesh and Eastern India. The findings clearly show that the double transplantation system is the appropriate technology for the early winter harvesting crop cultivation in the low lying areas to avoid the submergence problem in the flood-prone rice ecosystem in Bangladesh.

In Chapter 4; N.D. MacLeod, S. Wen and M. Hu examine the impact of the crop-cattle technology on improving economic conditions of households in China. Using data collected from smallholder households of six villages in Jiangxi and five villages in Hunan provinces in the Red Soils Region of Southern China, they have demonstrated how cattle-raising activities based on producing and feeding improved forages can potentially increase the economic welfare of smallholder households. The results from this Australian Centre for International Agricultural Research (ACIAR) sponsored project indicate that profits from cattle raising may significantly be increased with careful planning of forage production and management practices.

In the context of public agricultural research systems in both developed and developing countries being increasingly undermined in the sense of reduced funding, G.S. Ananth, P.G. Chengappa and

Aladas Janaiah highlight the impact of research investment on technology development and total factor productivity in major field crops of peninsular India. Using 25 years of time series data, their results show that agricultural research investment had considerable impact on release of crop varieties and other technologies in the state of Karnataka in India. The rate of return to agricultural research, which is estimated using the total factor productivity (TFP) approach, shows a high rate of return in the case of rice and sugarcane, and moderate for millet, cotton and sorghum. The growth in TFP was higher in the crops, which attracted higher research investments, in turn attributed to growth in yield due to continuous up-gradation of technologies.

In Chapter 6; Kei Kajisa has investigated the impact of dissemination of modern irrigation systems, that is, private wells with pumps, on the livelihood of, not only, the farmers who have access to wells but also, those farmers who have no access to wells and thus rely solely on traditional community-based irrigation systems called tank irrigation systems. The analysis is based on a village and household data set collected in Tamil Nadu, India, where tank irrigation systems have been managed collectively for rice cultivation. His statistical analyses predict that once the decline in collective management occurs due to the dissemination of private wells, the rice yield and income of the no-well-access farmers alone will decrease, resulting in increased poverty among them. His analyses also find that the dissemination leads to the overexploitation of groundwater and thus results in no significant increase in prof-itability of rice farming among the well-access farmers. In this way, the dissemination of private wells creates double tragedies: not only collapse of collective management of tank irrigation among the no-well-access farmers but also overexploitation and profit reduction among the well-access farmers in the state of Tamil Nadu in India.

In Chapter 7, Keijiro Otsuka, Jonna Estudillo, and Yasuyuki Sawada examine the impacts of the Green Revolution technology on income and schooling of children based on their own case studies of the three rice-growing villages in the Philippines, surveyed in 1985 and 2004. It is well known that the adoption of modern rice varieties has a significant positive impact on rice yield and income of rice farmers in irrigated areas. According to new evidence, the adoption of modern rice technology increased farm income positively even in rain-fed areas. This is likely because production efficiency in rain-fed ecosystems has been improved, due partly to the development of drought-tolerant varieties (shift of the frontier of the production function) and partly to the learning of improved management practices by rain-fed farmers. Such efficiency gain has significant long-term implications for the welfare of farmers, because schooling investment of children is affected by current income and educated children tend to find lucrative non-farm jobs in future.

In Chapter 5, Russell Wise and Oscar Cacho examine the impact of the crop-tree technology on mitigating climate change by sequestering carbon (C), which is also an alternative to arrest the land degradation emanating from shifting-cultivation and continuous cropping systems followed in Southeast Asia. In the policy context of the Kyoto Protocol (Article 3.3 and Article 12), Wise and Cacho have attempted to develop a meta-model of an agroforestry system and incorporate it into a dynamic programming (DP) algorithm to determine optimal management strategies in the presence of carbon (C) payments.

In Chapter 8, Suwanna Praneetvatakul and Hermann Waibel suggest a modelling framework to measure environment and economic impacts of farmer field school on crop and pest management practices of rice in Thailand. Using panel data from 241 farm households collected three times over a period of four years in five rice producing provinces of Thailand, the authors have shown that trained farmers significantly reduced pesticide use on the short-term and that they retained their practice of reducing pesticide use several years after the training.

References

Bhalla, G.S. and Peter Hazell. 1998.'Food Grains Demand in India to 2020—A Preliminary Exercise', *Economic and Political Weekly*, 32(52): pp. 150–64.

Hayami, Yujiro and Keijiro Otsuka. 1993. *The Economics of Contract Choice: An Agrarian Perspective*. Oxford, UK: Clarendon Press.

Hayami, Yujiro and Vernon W. Ruttan. 1985. *Agricultural Development: An International Perspective*. Baltimore, MD: Johns Hopkins University Press.

IRRI. 1979. *Farm Level Constraints to High Rice Yields in Asia: 1974–77*. Los Banos: International Rice Research Institute, Philippines.

———. 2006. *Bringing Hope, Improving Lives: Strategic Plan 2007–2015*. Los Banos: International Rice Research Institute, Philippines.

Janaiah, Aldas, Mahabub Hossain and Keijiro Otsuka. 2006. 'Productivity Impacts of the Modern Varieties of Rice in India', *Developing Economies*, XLIV-2 (June): pp. 190–207.

Kalirajan, Kaliappa. 1990. *Rice Production: An Econometric Analysis*. New Delhi: Oxford & IBH.

Otsuka, Keijiro and Frank Place. 2001. *Land Tenure and Natural Resource Management: A Comparative Study of Agrarian Communities in Asia and Africa*. Baltimore, MD: Johns Hopkins University Press.

Otsuka, Keijiro and Kaliappa Kalirajan. 2006. 'Rice Green Revolution in Asia and Its Transferability to Africa: An Introduction', *Developing Economies*, XLIV-2 (June): pp. 107–22.

2

Economic Assessment of an Indigenous Technology in Bangladesh

MD ABDUS SAMAD AZAD AND MAHABUB HOSSAIN

Introduction

Rice is the dominant staple food and source of energy intake of the 140 million people of Bangladesh. Rice occupies about 75 per cent of the total cropped area, contributes to over 50 per cent of the agricultural value added, and continues to be the main source of livelihood in rural Bangladesh (BBS, 2004). Aman rice (monsoon-season rice) is the traditional rice crop grown during the months of July to December in 5.5 million hectares of land and contributes to 45 per cent of the total rice production (BBS, 2005). In the greater Rangpur region (the northern part of Bangladesh), a large proportion of the land is medium and low-lying which is subjected to the risk of flooding from heavy rains during the months of August and September as is shown in Figure 2.2. Hence farmers have to transplant tall and aged seedling so that the Aman plant can tolerate this climatic stress. Most farmers practise a double transplantation system of crop establishment: transfer about 30 day-old seedlings from the seedbed to a relatively high level field with dense transplantation, and then transfer about 60 day-old seedlings to the main field when the seedlings are tall and the risk of flooding is over. This method of crop establishment is known locally as the *Bolon* system. It is an indigenous technology developed by farmers to manage the risk of submergence.

The objective of this chapter is to assess the productivity, profitability and technical efficiency of Aman rice cultivation under the *Bolon* system compared to the single transplantation system (locally called *Naicha*), commonly practised elsewhere in Bangladesh and Eastern India. As a background to the analysis, the *Bolon* system is explained in section II. The method of data collection and the analytical procedures are explained in section III. The findings of the survey on costs and returns and comparative technical efficiency are discussed in section IV. Section V draws implication for crop improvement research on *Bolon* for further refining the system.

The *Bolon* System

Bolon is a crop establishment system for rice cultivation where farmers transplant Aman rice seedlings twice, first on a piece of high land and then in the main field after the recession of heavy rains (Figure 2.1). The seedlings are prepared in a seedbed for about a month. They are first transplanted in a plot referred to as the *Bolon bari*. Farmers transplant seedlings in the *Bolon* plot with closer spacing and a large number of seedlings per hill. They take care of this plot like a main rice field and apply chemical fertilizers and insecticides to nurture the seedlings. After another 25–30 days (or when the risk of crop failure from heavy rains is less), these aged seedlings are again transplanted to the main field with broader spacing and less number of seedlings per hill. The seedlings grown in one decimal area of the seedbed (*Bechan bari*)

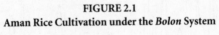

FIGURE 2.1
Aman Rice Cultivation under the *Bolon* System

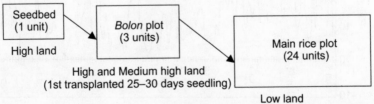

would occupy three decimals of land in the *Bolon bari*, and would be sufficient for transplanting about 24 decimals of land in the main rice field (*Dhan bari*). At the time of the final transplantation, the seedbed and the *Bolon bari* are also covered with sparse transplantation, so no land is wasted.

The *Bolon* system permits a flexible late transplantation during the rainy season, which is its prime advantage. The rice growers can transplant rice in their main plot/field at an advantageous time with regard to seasonal pattern of rainfall (Figure 2.2). The medium to low lying parcels of rice land are generally submerged during August to mid-September if there are consecutive days of heavy rainfall. So, this type of land is not suitable for transplanting young seedlings directly from the seedbed to main plot as the seedling is not tall enough and there is the risk of submergence. Under the *Bolon* system, aged tall seedlings are transplanted to the low-lying rice fields from the *Bolon* plot late in the season when the probability of occurrence of consecutive days of heavy rains is less; even if it occurs, the tall seedlings would not be submerged. Thus, this method of crop establishment, which has been developed by the farmers themselves and hence is an indigenous technology, helps avoid submergence of the rice plant from the uneven distribution of rainfall during the peak monsoon season.

FIGURE 2.2
Daily Average Rainfall Data, 1992–2003 in the Study Areas

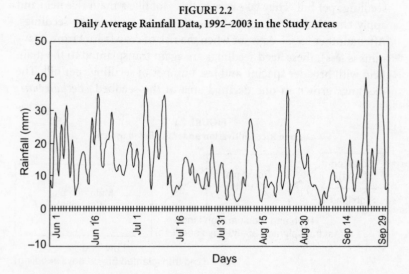

For medium to high lands, where there is little risk of ubmergence from heavy rains, farmers follow the single transplantation system early in the season (July to mid-August) with young seedlings directly pulled from the seedbed. This method is locally known as *Naicha* and is practised elsewhere in Bangladesh in lands of shallow flooding depth.

The biological scientists engaged in rice research in Bangladesh argue that the double transplantation system is an inefficient method of rice establishment compared to the single transplantation system. They argue that the aged seedlings would have less time to produce tillers in the field when compared to the young seedlings, and hence the crop stand would be less dense in the double transplanted field, resulting in lower crop yield. Further, the cost of rice production would be higher in the double transplanted systems due to the additional cost of labour for transplantation in the *Bolon* field, and for other aspects of crop care such as weeding and the application of fertilizers and pesticides, which can be avoided under the single transplantation system. Thus, the profitability of rice farming would be less under the *Bolon* system.

In a focus group meeting on the *Bolon* system that we organized before conducting the survey, the farmers refuted the argument of inefficiency of the double transplantation system compared to the single transplantation method. The farmers' experience is that the plants on double transplanted plots are usually healthy, have longer panicles and more filled grains than the plants on single transplanted parcels. They adjust the spacing at the time of transplanting (aged seedlings are transplanted more densely than younger seedlings) to ensure uniform crop stand under both systems. Through weeding and the use of pesticides in the *Bolon* plot (one-eighth the size of the main field) they avoid the need for further crop care in the main field and thereby reduce the cost of inter-cultural operations. According to the farmers, diseases and insect infestation is comparatively lower under this system than under single transplantation system. On the other hand, due to staggering of the time of transplantation (single transplanting on high land early in the season and double transplanting on low-land late in the season), farmers can avoid labour scarcity that they would face if all of them transplant rice at the same time after the onset of heavy rains. It helps utilizing family labour for longer periods and reduces the demand for hired labour, thereby putting less pressure on the labour market.

Therefore, the *Bolon* system appears to be an appropriate technology under the specific agro-ecological and environmental conditions. However, since the *Bolon* system involves additional cost for the second transplantation, it would be useful to estimate the magnitude of the cost and the associated benefits or costs in terms of the difference in yield and other crop management practices. The main purpose of the research is to generate data from a sample survey of the rice growers to study the economics of rice cultivation under *Bolon* system as compared to *Naicha*, which is the commonly practised crop establishment method for Aman rice throughout Bangladesh.

Methodology

To generate primary data, a farm household survey was conducted from a cross-section of farmers from five villages belonging to three northern districts—Rangpur, Lalmonirhat and Nilphamari—in 2004. The villages were purposively selected on the basis of prior knowledge of the *upazila* officials of the Department of Agricultural Extension (DAE) regarding the land type and the existence of the *Bolon* system. Therefore, three villages from Rangpur, one from Lalmonirhat and another one from Nilphamari district were finally selected. The villages were selected such that they have a large proportion of area under flooding depth of 50 cm to 100 cm during the peak of the monsoon season, while some other areas are under shallow flooding depth, so the farmers also practise the single transplantation system. For each of the selected villages, a list of farm households was drawn through a census of all households in the village and based on information on the method of establishment of the Aman crop. Then 40 farmers were selected randomly from the list of those farm households, who practised both the double transplantation and the single transplantation systems for the parcels operated by the farm. Size of the operating farm was also considered while selecting the sample farmers. Thus the sample consisted of 200 farmers, 40 each from five villages scattered throughout three districts.

Data was collected by administering a structured questionnaire which contained questions on the socioeconomic background of the farmers and the details of the costs and returns of Aman rice cultivation of two parcels; one for which the farmer practised the *Bolon* system and the other, the *Naicha* system. The questionnaire was pre-tested by the first author to assess its relevance for the local conditions, check the local terms of the measurement of land and output units, and to avoid any lead questions. The questionnaire was revised in response to the experience of pilot testing. Three local investigators were hired and trained for administering the survey. The first author supervised the data collection and was involved in checking the filled-in question-naires and editing and processing of data.

The paired t-test was performed to test the significance in mean differences of inputs and outputs of rice production between two systems for the same household. On the other hand, the Stochastic Frontier Production model was used to estimate the technical effi-ciency of rice farms under both *Bolon* and *Naicha* crop establishment systems. Several researchers (Aigner et al., 1977; Meeusen and Van den Broeck, 1977; Aly et al., 1987; Battese and Coelli, 1995; Fan et al., 1997) have used the Cobb-Douglas production function model to measure technical efficiency for industries and farms. They proposed the estimation of a Stochastic Frontier Production function, where noise is accounted for by adding a symmetric error term (u_i) to the non-negative term to provide:

$$Ln (Y_i) = f (X_i; \beta) + \varepsilon_i$$

Where, $\varepsilon_i = v_i - u_i$; $\qquad\qquad i = 1,\ldots\ldots, N$

Y_i denotes the output quantity of the i'th farm, X_i is a vector of the input quantities used by the farm, β is a vector of parameters to be estimated and ε_i is the error term composed of v and u. v_i is an independently and identically distributed random error with N $(0, \sigma v^2)$. These factors are outside the control of the firm. u_i is a non-negative random variable which is independently and identically distributed as N $(0, \sigma u^2)$, that is, the distribution of u_i is half normal. $| u_i | > 0$ reflects the technical efficiency relative to the frontier. $| u_i | = 0$ for a firm whose production lies on the frontier and $| u_i | < 0$ for a firm whose production lies below the frontier.

According to Battese and Coelli (1995), technical inefficiency effects are defined by:

$$u_i = z_i \delta + w_i, \qquad\qquad i = 1,\ldots\ldots, N$$

where z_i is a vector of explanatory variables associated with the technical inefficiency effects and δ is a vector of unknown parameter to be estimated. w_i is an unobservable random variable, which is assumed to be identically distributed, obtained by truncation of the normal distribution with mean zero and unknown variance σ^2, such that u_i is non-negative. According to Battese and Corra (1977), the variance ratio parameter γ which relates the variability of u_i to total variability (σ^2) can be calculated in the following manner:

$$\gamma = \sigma u^2 / \sigma^2, \qquad \text{where } \sigma^2 = \sigma u^2 + \sigma v^2; \qquad \text{so that } 0 \leq \gamma \leq 1$$

If the value of γ equals zero, the difference between farmers' yield and the efficient yield is entirely due to statistical noise. On the other hand, a value of one would indicate the difference attributed to the farmers' less than efficient use of technology, that is, technical inefficiency (Coelli, 1995).

The stochastic frontier production function can be estimated using either the maximum likelihood method or using a variant of the corrected ordinary least squares (COLS) method suggested by Richmond (1974). For this study, we have used the maximum likelihood method because of the availability of software; the Frontier Programme (Coelli, 1994) has automated the maximum likelihood method.

Kalirajan and Shand (1999) reviewed various methodologies for measuring technical efficiency and mentioned some limitations of stochastic frontier production function approach in their study. First, in the stochastic frontier production function approach, the level of individual technical inefficiency cannot be consistently estimated in a single cross-section. With a panel of data, this deficiency is resolved essentially because the noise is being averaged in the overall residual. Second, the farm potential is calculated by allowing technical efficiency to vary over all inputs taken together in this approach. Therefore, in the stochastic frontier approach the frontier parameter estimates cannot be identified with farms following the best practice techniques of applying inputs in the sample. While other methods such as Data Envelopment Analysis (DEA) and Stochastic Varying Coefficients Frontier Approach (SVFA) can facilitate identification of a benchmark of excellence in terms of the best practices in a given sample of observations, the stochastic frontier approach can only provide a signal to indicate whether a firm's overall performance is adequate in terms of realizing its own potential.

However, the following frontier production function model specifications were used in the study analysis:

$$\text{Ln } Y_i = \beta_0 + \beta_1 \text{LnX}_{1i} + \beta_2 \text{LnX}_{2i} + \beta_3 \text{LnX}_{3i} + \beta_4 D_1 + \beta_5 D_2 + \beta_6 D_3 + v_i - u_i$$

Where Ln denotes logarithms to base e

Y	=	Yield of rice (ton/ha)
X_1	=	Seed (Kg/ha)
X_2	=	Fertilizer (Kg/ha)
X_3	=	Human labour (man-days/ha)
D_1	=	Dummy crop establishment method (1 = *Bolon*, 0 = *Naicha*)
D_2	=	Dummy low land (1 = low land, 0 = otherwise)
D_3	=	Dummy high land (1 = high land, 0 = otherwise)
$v_i - u_i$	=	Decomposed error term as specified in Battese and Coelli (1995)

The technical inefficiency model can be expressed as:

$$u_i = \delta_0 + \delta_1 Z_1 + \delta_2 Z_2 + \delta_3 Z_3 + \delta_4 Z_4 + \delta_5 Z_5 + W_i \text{ where;}$$

Z_1 = Age of farmer (years)
Z_2 = Education (years of schooling)
Z_3 = Dummy farm size (1= farm size is greater than 1 hectare, 0 = otherwise)
Z_4 = Tenancy ratio
Z_5 = Dummy location (1= farmer resides in Rangpur, 0 = otherwise)
W_i = unobservable random variables

The explanatory variables are included in the above model because these variables may influence the level of technical efficiency of the agricultural production farms. Age is a variable included to estimate the impact of age on the level of technical efficiency. The variable education, that is, the number of years of schooling achieved by the farmer, is used as a proxy for marginal input. Kalirajan and Shand (1985) found in their study that education reduces technical inefficiency. Aragie (2001) also points out that education is an important positive determinant of a farmer's technical efficiency. The geographic location is the other farm characteristic that is used to account for any site specific factor (for example, soil fertility, differences in weather, and so on) not included in the production function but that may affect the level of farmers' technical efficiency. Aragie (2001) showed in his study that geographic location of the farm has an impact on farm efficiency. Likewise, tenancy ratio may influence the level of technical efficiency and hence it is included in the model.

Results and Discussion

Characteristics of Sample Farmers

The study analyzed the socioeconomic characteristics of the sample farmers of the survey areas and the findings are presented in Table 2.1. Most of the farmers (54 per cent) were young, between 20–35 years old. The average age of the sample farmers was 40 years. Although the literacy rate in Bangladesh is now increasing, most of the farmers had little education. About 35 per cent farmers did not have formal education, 24 per cent attended only primary schools and 17 per cent had secondary level education. Only 23 per cent of the sample farmers had completed higher schools or attended colleges.

The farmers under the study were classified into four groups based on their operational land holding. The average farm size of the sample farmers in all the study areas was found to be 0.90 hectare, which indicates that most of the farms operate at the subsistence level. The average farm size of the sample farmers is higher than the average for Bangladesh estimated at 0.67 hectare by the latest Agricultural Census taken in 1996 (BBS, 1999). According to the self-perception of the farmers, only 7 per cent were rich, 65 per cent were solvent, 24 per cent were poor and 4 per cent were extremely poor. A Poverty Monitoring Survey conducted in 2004 by the Bangladesh Bureau of Statistics estimated the head count index of poverty at 45 per cent for rural households, 35 per cent for owner farmers, 40 per cent for tenants, and 75 per cent for agricultural labour households (BBS, 2005).

Land Type and Adoption of the *Bolon* System

Farmers were asked to report the depth of flooding during the peak of the monsoon season for each parcel of land in their land portfolio.

TABLE 2.1
Socioeconomic Characteristics of the Sample Farmers

Variables	Study Areas				
	Rangpur	Lalmonirhat	Nilphamari	All	
Age of farmers (years)					
20–35 (young)	66	20	23	109	(54)
36–50 (middle age)	43	16	12	71	(36)
Above 50 (aged farmer)	11	04	05	20	(10)
Education					
No formal schooling	42	07	22	71	(36)
Primary attended	28	13	07	48	(24)
Secondary attended	22	07	05	34	(17)
Secondary passed	28	13	06	47	(23)
Farm size (hectare)					
Up to 0.40 (marginal)	36	12	12	60	(30)
0.41–1.00 (small)	48	16	16	80	(40)
1.01–2.00 (medium)	24	08	08	40	(20)
Above 2.00 (large)	12	04	04	20	(10)
Tenancy status					
Pure owner	66	25	27	118	(59)
Owner-tenant	38	10	08	56	(28)
Pure tenant	16	05	05	26	(13)
Economic condition					
Extremely poor	03	02	03	08	(04)
Poor	30	05	12	47	(24)
Solvent	76	31	23	130	(65)
Rich	11	02	02	15	(07)

Notes: 1. Figures in Table indicate number of sample farmers and figures in parentheses indicate percentage of farmers
2. Total number of sample farmers = 200
3. Rangpur = 120, Lalmonirhat = 40, Nilphamari = 40
4. Rangpur includes three upazilas, while Lalmonirhat and Nilphamari include one upazila each.

The high land was defined as those parcels which were flooded up to a depth of 20 cm, medium-level land as those flooded up to 50 cm, and the low lands as those flooded at over 50 cm depth. The estimate from the survey shows that about 23 per cent of the parcels operated by the sample farmers were high land, 52 per cent were medium land and the remaining 25 per cent were low land. The double transplantation

(*Bolon*) system was adopted in about 71 per cent of the total Aman rice crop; 80 per cent for the low land, 73 per cent for the medium land and 42 per cent for the high land. The low lands, in which the *Bolon* system was not used, were flooded at more than one metre depth where the traditional tall and low-yielding deep water rice varieties were grown. These varieties were directly seeded in the months of March and April, maintained as an upland crop for about 8–10 weeks and then grown along with the rising flood water with the onset of flooding beginning of July (Catling, 1992). The farmers reported that the *Bolon* system was previously used mainly for the medium and low lands, but of late the practice was spreading to the high land also.

Crop Management Aspects of *Bolon* and *Naicha* Systems

Farmers practise a closer line spacing (6 × 10 cm) and higher number of seedling per hill (9–16) for transplanting 25–34 days old seedlings in the *Bolon* plot (Table 2.2). During the period 15 July to 5 August, farmers uproot the tall seedlings from the *Bolon* plots and transplant 60–65 days old seedlings to the main fields. Farmers who practice the *Naicha* use different age of seedlings (35–45 days) depending on the type of land, that is, aged seedlings for low land and younger seedlings for medium and high lands as shown in Table 2.2. However, the average crop duration of *Bolon* is reported at 120 days, 10 days shorter than the *Naicha* system.

Inputs Use and Costs of Production

Farmers require less amount of seed under the *Bolon* system. As a result of transplantation into another field (the *Bolon* plot), tillering occurs in the plants and hence the number of healthy seedlings increases. The survey data show that farmers used 43 kg of seeds per hectare under the *Bolon* system compared to 68 kg of seeds for *Naicha*. The savings on account of seed, land preparation, transplantation and chemical fertilizers was estimated from the survey at US$ 19.25 per hectare under the *Bolon* compared to the *Naicha* system. Regarding the use of labour, the farmers required an additional 14 person-days/hectare for transplanting seedlings to the second transplantation that could be saved under the *Naicha* system. Results also reveal that weed and pest

TABLE 2.2
Variation in Agronomic Parameters Practised in Rice Cultivation
under *Bolon* and *Naicha* Systems

Agronomic parameters	Bolon System		Naicha System
	Bolon Plot	*Main Plot*	
Age of seedlings for transplanting in the rice field (days)	25–34	60–65	35–45
Seedlings per hill	9–16	3–4	4–6
Line spacing (cm)			
Line to line	10	15	20
Plant to plant	6	15	15
Crop duration (days)		120	130

infestation was comparatively low in the rice field cultivated under the *Bolon* system. Since tall seedlings are used and the fields remain waterlogged from the beginning of the transplantation, the weed pressure on the *Bolon* plot is minimal.

Economics of Rice Cultivation under *Bolon* and *Naicha* Systems

The productivity of Aman rice was comparatively higher under the *Bolon* system than that of *Naicha* for all types of land. The average yield of rice grown in low land under *Bolon* system was 4.04 t/ha, about 8 per cent higher than that of *Naicha* (Table 2.3). Besides, the yield of rice was 4.10 and 3.82 t/ha under *Bolon* and *Naicha* systems, respectively, for medium land. Results of paired t-test indicate that yield difference between *Bolon* and *Naicha* systems was significant at 1 per cent level (Table 2.4). Therefore, farmers earned almost 8 per cent higher gross returns from both low and medium rice land following the *Bolon* over the *Naicha* system. The study shows that per hectare cost of rice cultivation (US$ 208) under *Bolon* system was marginally higher than that of *Naicha* (US$ 204), but the difference was not statistically significant as can be seen in Table 2.4. The cost of labour under the *Bolon* system was significantly higher than that of *Naicha*, but other costs such as pesticide was significantly lower under *Bolon* system. On balance, the total cost of cultivation was not significantly different between the two systems.

TABLE 2.3
Economics of Aman Rice Cultivation under *Bolon* and *Naicha*
Systems in Northern Districts of Bangladesh

Particulars	Low Land		Medium Land		High Land	
	Bolon	Naicha	Bolon	Naicha	Bolon	Naicha
Rice yield (ton/ha)	4.04	3.75	4.10	3.82	3.68	3.57
Gross return (US$/ha)	459	424	468	435	419	407
Gross cost (US$/ha)	204	202	218	203	218	210
Net return (US$/ha)	255	223	250	233	201	197

TABLE 2.4
Difference in Yield and Other Production Parameters
between *Bolon* and *Naicha* Systems

Parameters	Bolon	Naicha	Percent Difference	t- value	Significance Level
Yield (ton/ha)	4.00	3.79	5.5	2.66	0.010
Labour cost (US$/ha)	107	104	3.2	2.08	0.041
Pesticide cost (US$/ha)	5.5	7.3	−24.7	−2.77	0.007
Gross cost (US$/ha)	208	204	1.9	1.57	0.121
Net return (US$/ha)	237	216	9.4	2.55	0.013

Though the cost difference between the *Bolon* system and *Naicha* was marginal, higher yields under the *Bolon* system provided more profits than under *Naicha*. The net return from Aman rice cultivation under the *Bolon* system was US$ 255/ha for the low land, about 14% higher than that for *Naicha* and it was 7% higher under the *Bolon* system for the medium land. For all land types the profitability gain was 9% more for the *Bolon* system than that of *Naicha*, which was statistically significant as shown in Table 2.4.

Technical Inefficiency: Results from Stochastic Frontier Model

The Maximum Likelihood (ML) estimates of the estimated stochastic frontier model are presented in Table 2.5. The estimated ML coefficients

TABLE 2.5
Maximum Likelihood Estimates of the Stochastic
Frontier Production Function

Variables	Parameter	Coefficients	Standard Error	t-ratio
Constant	β_0	7.2860	0.6141	11.86***
Seed	β_1	0.0244	0.1407	0.173
Fertilizer	β_2	0.0578	0.0275	2.101**
Human labour	β_3	0.1247	0.0437	2.852***
Crop establishment method (dummy)	β_4	0.1127	0.0666	1.693*
Dummy low land	β_5	−0.0348	0.0247	−1.410*
Dummy high land	β_6	−0.0956	0.0358	−2.667***
	σ^2	0.0938	0.0325	2.882***
	γ	0.7883	0.0851	9.265***
Log Likelihood		90.41		

Notes: *** Significant at 1% probability level.
** Significant at 5% probability level.
* Significant at 10% probability level.

of human labour and fertilizer are 0.125 and 0.058 respectively, which are statistically significantly different from zero. This indicates that increment of the inputs, human labour and fertilizer by 1 per cent will increase output by 0.12 and 0.06 per cent, respectively. The significant coefficient of the dummy for crop establishment method indicates that rice yield under the *Bolon* system is significantly higher than that of *Naicha*. The coefficients of dummy variables for both high and low lands are negative and statistically significant. These results indicate that technical efficiency in rice production is the highest for the medium-level land.

The estimated coefficients in the inefficiency model are presented in Table 2.6. The technical inefficiency is positively related with the age of the farmers, which indicates older farmers are less efficient than younger ones. The negative and significant coefficient for education shows that the educated farmers are more efficient than the non-educated farmers. The technical inefficiency is positively related with farm size, indicating that smaller farmers are more efficient than the larger ones. The coefficient of tenancy variable is negative and statistically significant, indicating that tenant farmers are more efficient than owner farmers.

TABLE 2.6
Determinants of Technical Inefficiency Model

Variables	Parameters	Coefficients	Standard-error	t-ratio
Constant	δ_0	−1.442	0.9993	−1.443*
Age of farmer	δ_1	0.012	0.0085	1.443*
Education	δ_2	−0.083	0.0598	−1.393*
Farm size (dummy)	δ_3	0.110	0.0718	1.531*
Tenancy ratio	δ_4	−0.440	0.3253	−1.354*
Location (dummy)	δ_5	0.322	0.3254	0.9895

Note: * Significant at 10% probability level.

The mean technical efficiency of the rice farms is found at 90 per cent for the *Bolon* system and 84 per cent for the *Naicha* (Table 2.7). The estimates show fairly high levels of efficiency in Aman rice cultivation in Bangladesh. As the technical efficiency of *Bolon* farms is higher than that of *Naicha*, it can be concluded that the *Bolon* system is technically more efficient than the *Naicha* system. The higher efficiency seems to be related to the farmers adapting appropriate crop establishment and management systems to the particular environment they face.

TABLE 2.7
Technical Efficiency Estimates of the Rice Farms
under *Bolon* and *Naicha* Systems

Crop Establishment System	Technical Efficiency Score	
	Mean	Standard Deviation
Bolon	0.90	0.021
Naicha	0.84	0.066

Farmers' Perceptions on *Bolon* System

The qualitative response obtained from the farmers indicated the following merits and demerits of the *Bolon* system:

Advantages of Bolon System

- Farmers can transplant rice plant in the field at a time when the risk of consecutive days of heavy rains is less.
- The aged and tall *Bolon* seedlings can be transplanted to the rice field with stagnant water.

- Healthy seedlings produce longer and uniform panicles resulting in less unfilled grains, higher grain weight and, therefore, higher productivity compared to the single transplantation system.
- Helps save seeds, pesticides and weeding labour.
- Greater opportunity for using family labour through phasing of planting dates for lands of different elevation and hence reduces dependence on hired labour.
- Ease labour scarcity and avoid high wage rate during transplantation.

Disadvantages of Bolon System

- Additional cost involved in land preparation and transplantation of seedlings at the *Bolon* plot.
- Bold rice straw produced from *Bolon* cultivation system is not good as cattle feed.
- This system is less convenient for the big farmers because many plots need to be transplanted twice.

Conclusions and Policy Implications

The findings of the study indicate that the *Bolon* system that farmers practise for Aman rice cultivation in the low-lying areas is an appropriate technology to avoid the problem of submergence in the flood-prone rice ecosystem. Higher productivity as well as significant net return from rice cultivation using the *Bolon* system indicate that farmers indeed gain by adopting this method of crop establishment. The stochastic frontier model also proved that rice farming under *Bolon* system is more efficient than that under *Naicha*. Due to the higher productivity and efficiency, the *Bolon* system has spread even to the medium and high lands where there is little risk of submergence from heavy rains. Therefore, instead of discarding the system, rice researchers must work on refining the system by developing appropriate varieties and other crop management practices for the system. Rice breeders at the International Rice Research Institute and the National Agricultural Research System have just developed submergence-tolerant high-yielding rice varieties for the flood-prone environment by incorporating the gene of local deep-water rice varieties with a high-yielding rice variety into a widely adopted rice variety (Swarna) in Eastern India. This discovery, coupled with the farmers' indigenous knowledge of crop management, may contribute to a substantial increase in rice production in the flood-prone ecosystem in Bangladesh and Eastern India.

References

Aigner, D.J., C.A.K. Lovell and P. Schmidt. 1977. 'Formulation and Estimation of Stochastic Frontier Production Function Models', *Journal of Econometrics*, 6: pp. 21–27.

Aly, H., Y. Belbase, K. Grabowski and S. Kraft. 1987. 'The Technical Efficiency of Illinois Grain Farms: An Application of a Ray-homothetic Production Function', *Agricultural Economics*, 19: pp. 69–78.

Aragie, K.K. June 2001. *Farm Household Technical Efficiency: A Stochastic Frontier Analysis.* A Study of Rice Producers in Mardi Watershed in the Western Development Region of Nepal. A Master Thesis Submitted to the Department of Economics and Social Sciences, Agricultural University of Norway.

Bangladesh Bureau of Statistics (BBS). 1999. *Census of Agriculture 1996.* Vol. 1, Statistics Division, Ministry of Planning, Government of the People's Republic of Bangladesh.

————. 2004. *Monthly Statistical Bulletin August 2004*, Statistics Division, Ministry of Planning, Government of the People's Republic of Bangladesh.

————. 2005. *Report of the Poverty Monitoring Survey 2004*, Statistics Division, Ministry of Planning, Government of the People's Republic of Bangladesh.

Battese, G.E. and G.S. Corra. 1977. 'Estimation of a Frontier Model: With Application to the Pastoral Zone of Eastern Australia', *Australian Journal of Agricultural Economics*, 21: pp. 167–79.

Battese, G.E. and T.J. Coelli. 1995. 'A Model for Technical Inefficiency in a Stochastic Frontier Production Function for Panel Data', *Empirical Economics*, 20: pp. 325–32.

Catling, D. 1992. *Rice in Deep Water*, International Rice Research Institute, Los Baños, Philippines.

Coelli, T.J. 1994. *A Guide to Frontier Version 4.1: A Computer Program for Stochastic Frontier Production and Cost Function Estimation*, Department of Econometrics, University of New England, Australia.

————. 1995. 'Recent Developments in Frontier Modeling and Efficiency Measurement', *Australian Journal of Agricultural Economics*, 39(3): pp. 219–45.

Fan, S.F., E.J. Wailes and K.B. Young. 1997. 'Policy Reforms and Technological Change in Egyptian Rice Production: A Frontier Production Function Approach', *Journal of African Economics*, 6(3): pp. 391–411.

Kalirajan, K.P. and R.T. Shand. 1985. 'Types of Education and Agricultural Productivity: A Quantitative Analysis of Tamil Rice Farming', *Journal of Devel-opment Studies*, 21: pp. 222–243.

————. 1999. 'Frontier Production Functions and Technical Efficiency Measures', *Journal of Economic Surveys*, 13(2): pp. 149–72.

Meeusen, W. and J. Van den Broeck. 1977. 'Efficiency Estimation from Cobb-Douglas Production Functions with Composed Error', *International Economic Review*, 18: pp. 435–44.

Richmond, J. 1974. 'Estimating the Efficiency of Production', *International Economic Review*, 15: pp. 515–21.

3

Research Investment on Technology Development in Peninsular India

G.S. ANANTH, P.G. CHENGAPPA AND ALDAS JANAIAH

Introduction

Productivity growth is of paramount importance for economic growth. Increase in agriculture productivity can be induced by public investments in research extension, human capital formation and infrastructure (Rosegrant and Evenson, 1994). Public investments in infrastructure, research and extension along with crop production strategies have helped to expand crop production and grain stocks in India. Of late, the investment in agriculture, particularly on agricultural research and development, is on the decline so that future growth in production can only be input-based in many regions of the country (Kumar and Rosegrant, 1994). Public agricultural research systems in both developed and developing countries are increasingly facing resource crunch and are lately seeking priority-setting procedures to help mediate the often-conflicting demands (Norton et al., 1992).

There are several studies carried out to quantify pay-offs to agricultural research investment at aggregate (all India) level (Evenson and Jha, 1973; Evenson and McKinsey, 1991; Kumar and Rosegrant, 1994; Coelli and Rao, 2005). In India, agriculture is a state subject with the Central (Federal) Government playing the supportive role. The agriculture sector has been the focal subject of State Governments as more than two-thirds of the population is engaged in this sector. All states have invested considerable amount of resources to promote agricultural research, extension and education. But, hardly any attempts have been made to quantify returns to research investment in agriculture at the state or regional level. In this context, an in-depth analysis of returns to research investment at the state level assumes a greater significance. The specific objectives of the study are to measure (a) the impact of research investment on the development of technologies, and (b) the rate of return to research investment made in the state.

Data Sources and Methods of Analysis

The present study is an attempt to quantify the returns to research investment on major field crops namely, rice, jowar, ragi, red gram, groundnut, sunflower, cotton and sugarcane grown in different agro-climatic zones of Karnataka, a predominantly agricultural state in peninsular India. These crops together occupy nearly 65 per cent of the gross cropped area and account for a larger chunk (43 per cent) of the research investment in the state. The secondary data on expenditure incurred on research at the two State Agricultural Universities (SAUs) in Karnataka was collected for a period of 25 years from 1970–71 to 1994–95. The data on technology and crop varieties released have been compiled from the annual reports of both the SAUs for the period 1970–71 to 1994–95. Information such as the district-wise area, production and productivity of major field crops was collected for the period 1970–71 to 1994–95 from the reports published by the Directorate of Economics and Statistics, Government of Karnataka. The data on cost and returns of major field crops 1980–81 to 1994–95 was collected from the Farm Management Survey reports published by the Directorate of Agriculture, Government of Karnataka. The data on the prices of inputs and agricultural commodities have been collected from various secondary sources for the period 1970–71 to 1994–95 in order to construct the indices.

The impact of agricultural research investment can be captured by adopting different approaches; production function, rate of growth, total factor productivity (TFP), co-integration and distributed lag model. Of these, the TFP approach is more comprehensive in which the research contribution can be specifically captured. The TFP concept, which implies an index of output per unit of total factor inputs, measures these shifts in output appropriately, holding all inputs constant (Kumar and Mruthunjaya, 1992).

The crop varieties and technology relating to non-varieties developed and released by the two SAUs for major field crops were cross tabulated for three periods namely, 1970–71 to 1979–80; 1980–81 to 1989–90 and 1990–91 to 1994–95 to assess the period and discipline-wise performance.

Input Saving Effect of Technology

A simple way to assess the impact of technology on the state's agriculture is to estimate the amount of land saved due to yield-augmenting technology in the form of improved productivity realized by the crops. Land saved due to productivity improvement was calculated by taking the difference between the land required to produce the current level of production with the technology that prevailed during the base or reference period and the actual land being used presently (Bisaliah and Patil, 1987).

The formula used is: $LS = \dfrac{P_1}{Y_0} - A_1$

Where,

LS = Land Saving (in million ha)

P_1 = the average production in latest five year period, that is, 1990–91 to 1994–95 (in million tonnes)

A_1 = the average area under the crop in the latest five year period, that is, 1990–91 to 1994–95 (in million hectares)

Y_0 = average yield in base five year period, that is, 1970–71 to 1974–75 (in tonnes/ha)

The TFP is the impact of input cost on output due to technological advancement in the crop sector. TFP is the ratio of total output index to the total input index or the value of total output to the total input cost multiplied by 100 or expressed in percentage over the years. The TFP concept which implies an index of output per unit of total inputs measures shifts in output appropriately, holding all inputs constant. Thus, TFP measures the amount of increase in total output, which is not accounted for, by the increase in total inputs (Kumar and Rosegrant, 1994). Farm harvest prices were used to aggregate the value of outputs. The inputs included in the input index were human labour, bullock labour, seed, manure, fertilizers, pesticides, irrigation and depreciation. The inputs were aggregated using the factor shares with appropriate weights. Total output, total input, total factor productivity indices were calculated as follows.

Total output index (TOI) =

$$TOI_t / TOI\ t-1 = \pi_j \left(\frac{Qj_t}{Qj_{t-1}} \right)^{(Rj_t + Rj_{t-1})^{1/2}}$$

Total input index (TII_t) =

$$TII_t / TII\ t-1 = \pi_i \left(\frac{Xi_t}{Xi_{t-1}} \right)^{(Si_t + SI_{t-1})^{1/2}}$$

Total factor productivity index (TFPI) =

$$TFPI_t = \left(\frac{TOI_t}{TII_t} \right) \times 100$$

Where,

R_{jt} = the share of output 'j' in total revenue
Q_{jt} = output 'j'
S_{it} = the share of input 'i' in total input cost
X_{it} = input 'i'

By specifying TOIt – 1 and TIIt – 1 equal to 100 in the initial year, the above equations provide the total output, total input and total factor productivity indices for the specified period 't'.

Returns to Research Investment

Changes in output other than those generated by changes in inputs can be induced by research, extension, human capital, infrastructure, price policy and climatic factors. As an input into public investment decisions, it is useful to understand the relative importance of productivity enhancing factors in determining productivity growth. In order to assess the determinants of TFP, the TFP index was regressed on crop research investment per hectare of area per year, which is a linear trend variable. The time series data from different years were pooled. Estimation was done by using a fixed-effects approach for the pooled cross section and time series data set. Using the elasticity of TFP with respect to research investment, one can easily estimate the value of marginal product (additional product value) of research investment (R) as:

$$VMP\ (R) = b \times (V/R)$$

Where,

R = the research investment
V = the value of the production associated with TFP
b = the TFP elasticity of research investment

Estimated in the TFP determinant equation above, the benefit stream generated under the assumption that the benefit of investment made in research in period $(t - 1)$ will start generating a benefit after a lag of years at an increasing rate in the beginning, remain constant for a period of time and thereafter decline following the typical inverted 'V' shape curve. A rupee invested in year $(t-1)$ will generate a benefit equal to 0.1 VMP in year $(t + 1)$ and so on. The rate of return to investment was obtained by using the discounted cash flow measure of the internal rate of return (IRR).

Research contribution/incremental production for year i =

$$\frac{TFPi \times Current\ year\ production}{100} - Base\ year\ production$$

Internal rate of return (IRR) = $\sum_{t=1}^{N} \frac{B_t - C_t}{(1+r)^t} = 0$

Where,

B_t = Benefit in year 't'
C_t = Cost in year 't'
r = Internal rate of return and is considered as the marginal rate of return to public research investment.

Results and Discussion

Impact of Research Investment on Technology Development

The two SAUs in Karnataka have developed and released 90 crop varieties in eight major field crops considered for the analysis during the period 1970–71 to 1994–95 as detailed in Appendix 3.1. Of these, cereals such as rice, ragi and jowar together accounted for 57 per cent of total varieties developed and released. During the first period (1970–71 to 1979–80) 49 varieties were released, of which cereals accounted for 63 per cent. During the first period, no variety was released in the case of red gram and fewer varieties were released in the case of groundnut, sunflower and sugarcane. During the third period, a total of 10 varieties were released, of which rice accounted for five and jowar and ragi two each. In addition to release of crop varieties, the focus of research has been to evolve appropriate location specific production technologies in order to realize the full potential of the crop in terms of productivity. During the period of analysis (1970–71 to 1994–95), 3,437 technologies pertaining nine disciplines have been released (Appendix 3.2). Of the total 3,437 technologies released, 1,144 have been released during the first period, 1,413 during the second period and 880 during the third period. In all the three periods, food grains comprising of rice, jowar, ragi and red gram accounted for the major share (60 per cent) of technologies released followed by commercial crops. This is under-standable since the state and the nation as a whole had placed emphasis on achieving self-sufficiency in food grain production. From the above, it is clear that the development and release of number of crop varieties is on the decline. However, there is a steady increase in the release of technologies other than these varieties, more so in cereals as compared to other crops over a period of time.

Discipline-wise Technologies Released in Major Field Crops in Karnataka

In both the SAUs, nine major disciplines were associated in evolving 3,237 technologies other than crop varieties in eight major field crops of the state technologies (Appendix 3.3). Of them, Agronomy and Genetics and Plant Breeding accounted for 57 per cent of the technologies released across the eight major field crops. The disciplines of Entomology and Plant Pathology accounted for 17 per cent with similar emphasis in Crop Physiology and Soil Science.

The agricultural research investment had considerable impact on release of crop varieties and other technologies, which led to improvement in productivity of the crops in the state during the period of analysis.

Impact of Research Investment on Input Saving in Major Field Crops

The yield augmenting technologies namely, development and release of high yielding crop varieties, agronomic practices and other crop improvement and plant protection technologies have had a considerable impact on saving of input use, such as land in the state, over the years. The land-saving effect due to higher productivity achieved through technologies can be observed from Table 3.1.

During the period 1990–91 to 1994–95, Karnataka produced an average of 4.09 million tonnes of rice in 1.28 million hectares of land. If, this 4.09 million tonnes of rice were to be produced at 1970–71 to 1974–75 yield level of 1.816 tonnes per hectare, Karnataka would have required 2.25 million hectares of land. This implies that Karnataka could save 0.096 million hectares (83.48 per cent) of land due to higher productivity achieved through technology for rice alone in the state over the 25-year period. Similar analysis for other crops indicated that land-saving effect was highest in the case of cotton (0.616 million hectares) followed by rice (0.096 million hectares), ragi (0.052 million hectares), groundnut (0.017 million hectares) and jowar (0.014 million hectares). There was very little land saving in the case of sugarcane (0.0003 million hectares). However, there was land loss seen in case of sunflower (–0.023 million hectares) and red gram (–0.015 million hectares). Hence, it is imperative that much emphasis be placed on research to evolve new technologies to increase the productivity of sunflower and red gram in the state so that the states can move towards high-value crops and thus increase their agricultural income.

TABLE 3.1
Land Saving Effect Due to Technology in Major Field Crops of Karnataka
(1970–71 to 1994–95)

Sl. No.	Crops		Average for the Years		Land Saving (in Million Hectare)	Percent Land Saved
			1970–71 to 1974–75	1990–91 to 1994–95		
1.	Rice	P	2.03	4.09		
		A	1.14	1.28	0.96	83.98
		Y	1816	3336		
2.	Jowar	P	0.16	1.67		
		A	2.29	2.16	0.14	5.92
		Y	734	817		
3.	Ragi	P	0.88	1.38		
		A	0.10	0.10	0.52	49.61
		Y	885	1405		
4.	Red gram	P	0.15	0.14		
		A	0.24	0.42	−0.15	−51.06
		Y	534	369		
5.	Groundnut	P	0.68	1.04		
		A	0.98	1.26	0.17	17.20
		Y	732	869		
			(1980–81 to 1984–85)			
6.	Sunflower	P	0.11	0.43		
		A	0.22	1.09	−0.23	104.95
		Y	501	420		
7.	Cotton	P	0.67			
		A	1.10	0.81		
		Y	115	0.61	6.16	561.08
				240		
8.	Sugarcane	P	9.06	25.38		
		A	0.11	0.29	0.003	2.73
		Y	86000	91000		

Notes: P = Production in million tonnes.

      A = Area in million hectares.

      Y = Productivity (yield) in kg per ha.

Impact of Research Investment on Rate of Return in Major Field Crops in Karnataka

The rate of return to agricultural research was estimated by using the total factor productivity (TFP) analysis. For the period 1980–81 to 1994–95, the increase in productivity attributable mainly to the research effort has been quantified through the total factor productivity index. This, in turn, was applied to the total area of crops to obtain the gross benefit of the research endeavour. The analysis was carried out for three sub-periods, namely, pre-stagnation of productivity (1981–82 to 1985–86), stagnation of productivity (1986–87 to 1990–91) and post-stagnation of productivity (1991–92 to 1994–95). These three sub-periods were identified based on the Report of the Expert Committee, Government of Karnataka (1993) on stagnation of Agricultural Productivity in Karnataka during the 1980s. Total factor productivity of major field crops in Karnataka has been calculated for the above three periods and the results are presented in Appendix Table 3.1. During the pre-stagnation period, TFP growth was over 1 per cent per annum in case of ragi, rice and cotton; about 1 per cent per annum in case of jowar and sugarcane and much lower in case of sunflower, groundnut and red gram. During the stagnation period, TFP growth was more than 1.3 per cent per annum in case of sugarcane and rice and about one per cent in case of cotton, jowar, and sunflower. The TFP growth was as low as 0.6 to 0.8 per cent per annum in the case of other crops, namely, groundnut, ragi and red gram crops in the state. During the post-stagnation period, the highest rate of 1.3 per cent growth per annum in TFP was observed in case of sugarcane and rice but it was 1 per cent in jowar crop. It is interesting to note that for the rest of the crops, namely, ragi, red gram, groundnut, sunflower and cotton, the TFP growth ranged between 0.4 and 0.6 per cent per annum. This indicates that TFP growth declined during post-stagnation period in the state. The TFP growth was around 1 per cent during the three periods in the case of rice, jowar, cotton and sugarcane, but less than one per cent in the case of red gram, groundnut, ragi and sunflower in the state as a whole. The decline in the TFP as well as low growth in TFP in different crops in different periods is explained by the fact that there is no commensurate increase in productivity, particularly in the case of red gram and sunflower, and due to a marginal increase in groundnut over the period. During the same period, the area under the crops also increased, barring the marginal decline in case of ragi and jowar. Moreover, it is observed that there has been an increase in the use of purchased inputs like fertilizers and hired labour, which resulted in increase in cost of cultivation even under rain-fed conditions. However, the increase in productivity in most of the crops under analysis is not in tune with the proportionate increase in cost resulting in lowering of TFP.

The rate of return (IRR) to research investment worked out for each crop indicates that it was the highest in case of rice (251 per cent) followed by sugarcane (59.82 per cent), cotton (21.68 per cent), ragi (16.79 per cent) and jowar (11.03 per cent). The returns were negative in respect of red gram, groundnut and sunflower. If the present declining trend in TFP observed in most crops is not arrested, the rate of return to research investment in future will be much lower. This implies that there is a need for achieving higher yield levels in the case of jowar, ragi, red gram, groundnut and sunflower through remandating the research and extension strategies. This is possible by increasing the investment in research to develop new technologies that will contribute to yield enhancement.

Conclusions

This chapter presents the impact of research investment in terms of technology development and rate of return based on 25 years of data relating to eight major field crops in peninsular India. The results show that agricultural research investment had a considerable impact on release of crop varieties and other technologies. Food grains comprising of rice, jowar, ragi and red gram accounted for a major share of technologies released followed by commercial crops. This is understandable since the state and the nation as a whole had placed emphasis on achieving self-sufficiency in food grain production and as a result more research investments were made in these crops. The land-saving effect due to yield-augmentation technologies such as high-yielding crop varieties, agronomic practices and plant protection was perceptible in crops such as cotton, rice, ragi and jowar. It was negative in case of sunflower and red gram due to a fall in yield levels, which calls for research investments to develop suitable high yielding varieties. The rate of return to agricultural research estimated using the total factor productivity shows high rate of return in the case of rice and sugarcane, moderate for finger millet, cotton and jowar and negative in red gram, groundnut and sunflower. The growth in TFP was also higher in the crops which attracted higher research investments, in turn attributed to growth in yield due to continuous upgradation of technologies. The state has achieved self-sufficiency in food grains but to meet the nutritional needs, it is imperative that the investments in pulses and oil seeds are augmented while continuing the present trend in the case of cereals and sugarcane.

APPENDIX TABLE 3.1

Total Factor Productivity Growth of Major Crops in Karnataka State

Sl. No.	Crops	Period	Average Total Area (million ha)	Average Total Production (million tonnes)	Average Productivity (kg/ha)	Average Total Research Expenditure (million rupees)	Average Research Expenditure (rupees/ha)	Additional Value (million rupees)	Average TFP growth	Internal Rate of Return (IRR %)
1.	Rice	Pre-stagnation	1.16	2.23	2035	5.15	4.47	1.80	101.54	
		Stagnation	1.16	2.54	2292	8.12	6.96	15.77	132.10	
		Post-stagnation	1.32	4.20	3354	14.74	11.20	23.22	133.35	251.00
2.	Jowar	Pre-stagnation	2.26	1.63	764	4.39	1.94	3.59	98.51	
		Stagnation	2.34	1.59	714	6.51	2.88	6.82	109.40	
		Post-stagnation	2.17	1.78	864	15.25	7.06	5.46	101.21	11.03
3.	Ragi	Pre-stagnation	1.10	1.23	1176	1.94	1.77	6.63	125.91	
		Stagnation	1.13	1.20	1112	4.22	3.76	4.98	74.42	
		Post-stagnation	1.03	1.48	1517	8.68	8.49	15.83	44.06	16.79
4.	Red gram	Pre-stagnation	0.39	0.18	473	0.91	2.32	0.77	88.81	
		Stagnation	0.47	0.18	409	1.41	3.71	1.39	80.73	
		Post-stagnation	0.41	0.14	385	2.51	8.41	4.45	56.89	Negative
5.	Groundnut	Pre-stagnation	0.92	0.70	798	1.21	1.32	3.31	81.39	
		Stagnation	1.16	0.87	795	2.65	2.26	9.57	61.82	
		Post-stagnation	1.27	1.10	909	5.67	4.52	19.78	52.15	Negative

No.	Crop	Period								
6.	Sunflower	Pre-stagnation	0.29	0.14	498	1.46	5.96	–	79.23	
		Stagnation	0.72	0.28	414	4.14	6.17	0.63	108.94	Negative
		Post-stagnation	1.15	0.45	413	7.47	6.94	0.54	59.10	
7.	Cotton	Pre-stagnation	0.88	0.65	134	3.20	3.80	4.91	110.88	
		Stagnation	0.57	0.69	219	4.62	8.02	8.86	118.62	14.68
		Post-stagnation	0.61	0.85	252	10.84	17.86	25.65	48.70	
8.	Sugar cane	Pre-stagnation	0.18	13.72	82	2.40	13.65	31.90	97.47	
		Stagnation	0.23	18.55	84	4.07	17.07	172.80	141.87	59.82
		Post-stagnation	0.30	26.54	93	9.73	32.73	224.33	134.48	

Notes: Pre-stagnation period (1981–82 to 1985–86); Stagnation period (1986–87 to 1990–91); Post-stagnation period (1991–92 to 1994–95); Productivity in column 6 is in tonnes per hectare in the case of sugarcane.

APPENDIX 3.2
Number of Varieties Developed and Released in Major Crops by SAUs in
Karnataka (1970–71 to 1994–95)

Sl. No.	Crops	Period I (1970–71 to 1979–80)	Period II (1980–81 to 1989–90)	Period III (1990–91 to 1994–95)	Overall Period (1970–71 to 1994–95)	Per cent to Total
1.	Rice	14	9	5	28	31.11
2.	Jowar	4	6	2	12	13.33
3.	Ragi	10	5	2	17	18.89
4.	Red gram	–	1	–	1	1.11
5.	Groundnut	4	1	–	5	5.56
6.	Sunflower	2	–	1	3	3.33
7.	Cotton	10	6	–	16	17.78
8.	Sugarcane	5	3	–	8	8.89
	Total	49	31	10	90	100

Source: Annual Reports of SAUs; UAS, Bangalore and UAS, Dharwad.

APPENDIX 3.3
Number of Technologies, Other than Varieties Developed and Released in Major
Field Crops by SAUs in Karnataka (1970–71 to 1994–95)

Sl. No.	Crops	Period I (1970–71 to 1979–80)	Period II (1980–81 to 1989–90)	Period III (1990–91 to 1994–95)	Overall Period (1970–71 to 1994–95)	Per cent to Total
1.	Rice	255	329	193	777	22.60
2.	Jowar	196	219	121	536	15.60
3.	Ragi	164	161	91	416	12.51
4.	Red gram	66	155	108	329	12.19
5.	Groundnut	144	164	122	430	12.10
6.	Sunflower	97	154	97	348	10.13
7.	Cotton	161	155	103	419	9.57
8.	Sugarcane	61	76	45	182	5.30
	Total	1144	1413	880	3437	100

Source: Annual Reports of SAUs; UAS, Bangalore and UAS, Dharwad.

APPENDIX 3.4

Discipline-wise Number of Technologies, Other than Crop Varieties Developed and Released in Major Field Crops by the SAUs of Karnataka (1970–71 to 1994–95)

Sl. No.	Discipline	Overall Period (1970–71 to 1994–95)									
		Rice	Jowar	Ragi	Red gram	Groundnut	Sunflower	Cotton	Sugar Cane	Total	%
1.	Agronomy	234	195	129	128	127	129	169	105	1216	37.56
2.	Genetics and Plant Breeding	29	73	125	70	99	81	93	52	622	19.22
3.	Crop Physiology	38	42	40	11	33	44	35	–	243	7.51
4.	Entomology	82	72	10	40	45	10	55	11	325	10.04
5.	Plant Pathology	59	46	56	21	37	25	25	7	276	8.53
6.	Agricultural Microbiology	25	9	13	36	20	6	10	5	124	3.83
7.	Agricultural Engineering	20	15	11	3	21	12	4	–	86	2.66
8.	Soil Science and Agricultural Chemistry	69	57	27	15	37	23	22	2	252	7.78
9.	Seed Technology	21	27	5	5	11	18	6	–	93	2.87
	Total	577	536	416	329	430	348	419	182	3237	100

Source: Annual reports of SAUs; UAS, Bangalore and Dharwad.

References

Bisaliah, S. and S.V. Patil. 1987. 'Technological Base and Agricultural Transformation of Karnataka', *Agricultural Situation in India*, 42 (5): pp. 367–72.

Coelli, Tim J. and D.S. Prasad Rao. 2005. 'Total Factor Productivity Growth in Agriculture: A Malmquist Index Analysis of 93 Countries', 1980–2000, Proceedings of the 25th, *International Conference of Agricultural Economists*, pp. 115–34.

Evenson, R.E. and Dayanath Jha. 1973. 'The Contribution of Agricultural Research System to Agricultural Production in India', *Indian Journal of Agricultural Economics*, 28(4): pp. 212–30.

Evenson, R.E. and Mckinsey W. James. 1991. 'Research, Extension, Infrastructure and Productivity Change in Indian Agriculture', in R.E. Evenson and C.E. Pray (eds), *Research and Productivity in Asian Agriculture*, pp. 155–83. Ithaca, N.Y.: Cornell University Press.

Government of Karnataka. 1993. 'Stagnation of Agricultural Productivity in Karnataka During 1980s', *Report of the Expert Committee*, Government of Karnataka, Bangalore.

Kumar, P. and Mruthunjaya. 1992. 'Measurement and Analysis of Total Factor Productivity Growth in Wheat', *Indian Journal of Agricultural Economics*, 47 (3): pp. 451–57.

Kumar, P. and W. Mark Rosegrant. 1994. 'Productivity and Sources of Growth for Rice in India', *Economic and Political Weekly*, 29 (53): pp. 183–88.

Norton, George, W. Philip, G. Pardey and Julian M. Alston. 1992. 'Economic Issues in Agricultural Research Priority', *American Journal of Agricultural Economics*, 74 (1): pp. 38–45.

Rosegrant W. Mark and R.E. Evenson. 1994. 'Total Factor Product of Long-term Growth in Indian Agriculture', Paper presented at IFPRI/Indian Agricultural Research Institute Workshop on Agricultural Growth in India, New Delhi, India.

4

Impact of Forage Options for Beef Production on Small Farms in China

N.D. MACLEOD, S. WEN AND M. HU

Introduction

Concurrent with rapid social, economic and industrial change in the People's Republic of China, in recent decades, the agricultural sector has experienced an unparalleled rate of growth. For example, while beef presently represents only about 8 per cent of the total output of meat, it expanded by almost 1,100 per cent, during the 15-year period to 2000, from almost negligible levels (Longworth et al. 2001). Despite the emergence of specialized feedlots (for example 50+ cattle on feed), beef production is largely oriented to low value local markets and is dominated by small-holder production units where meat production has traditionally been a secondary consideration to draught use. Cattle husbandry and nutrition on these small holdings is typically poor, with most of the diet based on low quality crop residues and limited, tethered or free grazing of fallowed land, roadsides, terraces and wastelands (Xie 1991). In more recent years, public policy for further promoting beef production has been directed to the use of planted forages and energy dense or protein-augmented crop residues to improve diet quality and raise productivity (Xie et al. 1992; Clements et al. 1997).

The Red Soils Region of southern China (Figure 4.1), encompassing 2.6 million km^2 of land across 14 provinces, was identified in the

FIGURE 4.1
Soil Map of South China (after Xi et al., 1990)

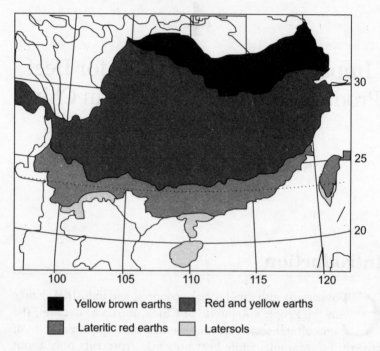

■ Yellow brown earths	■ Red and yellow earths
▨ Lateritic red earths	▢ Latersols

Source: Xi, C.F., Q. Xu and J. Zhang. 1990. 'Distribution and Regionalisation of Soils'. In *Soils of China*, Institute of Soil Science, Science Press, Beijing, pp. 20–39.

late 1980's as an area in which expanded beef production based on forages might be successfully encouraged (Sturgeon 1991). A series of agronomic projects[1], sponsored by the Australian Centre for International Agricultural Research (ACIAR) in partnership with the Chinese Government through the 1990's, were collectively successful in identifying forage species that are generally persistent and productive in the region; identifying fertilizer regimes to correct the major soil nutrient deficiencies; and investigating the palatability of these species to ruminants (Clements et al. 1997).

In 2000, ACIAR in partnership with the Chinese Academy of Agricultural Sciences and the Jiangxi Ministry of Science and Technology, sponsored a new project[2] to integrate the outputs of the previous research through a series of trials that covered cattle feeding, the growing and handling of forages, residues and energy-dense supplements, and the definition of 'feed year' plans for the Red Soils Region. The work was conducted at two sites—Nanchang, Jiangxi Province and Qiyang, Hunan Province, and is being further promoted through the establishment of demonstration farms by local offices of the Animal Husbandry Bureau in a number of counties in Jiangxi and Hunan Provinces. This project also sought to provide an economic assessment of the prospective benefit of incorporating forage-based cattle-feeding systems to small-holder households within their existing farming systems.

This chapter presents an assessment, based on modelling and a synthetic case study, of the prospective impact on the profitability of a small-holder production enterprise that incorporates a modest cattle finishing activity, based on the use of farm-grown forages, crop residues and purchased supplements, within its existing cropping system. The feeding regime is based on a feed year plan that is consistent with the ACIAR project recommendations and the level of animal performance is based on the mean results for the cattle feeding trials. The analysis includes a simple sensitivity test of changes in two key variables, namely, finished stock prices and live-weight gain. Some implications of the results for the farmer demonstration programmes are presented.

Material and Methods

The ACIAR Project seeks to develop economically viable forage-based beef production systems in the Red Soils Region. Published data on the productivity and economic performance of individual small-holder farms within the Red Soils Region with forage and beef activities are essentially non-existent. Therefore, the prospective impact of forage production and feeding strategies for beef production activities on small-holder farms is illustrated using a synthetic farm modelling approach. A case study is presented for a typical small-holder farm household assumed to be located in Chunqiutang Village, Hunan Province, whose structure is detailed in a following sub-section.

Household Surveys

A survey was conducted in four counties within Jiangxi and Hunan Provinces to provide baseline data for small-holder enterprises, and to calibrate an economic model of a small-holder enterprise. The counties are Taihe and Gao'an Counties (Jiangxi) and Dao and Jiang Yong Counties (Hunan), and these were identified as having a strong interest in promoting economic development through increased beef production. Small-holder households were surveyed in six villages in Jiangxi by students from Jiangxi Agricultural University and in five villages in Hunan by students from Hunan Agricultural University during October and November 2002. While the surveying of 10 of the 11 villages was based on random sampling of households (sample size 20–30 per cent), the survey of Chunqiutang village in Dao County, Hunan, involved a census of all households ($n = N = 135$)[3].

Detailed results from the village surveys are included in the final project report (MacLeod et al. 2004). To demonstrate the application of the economic model, only the data for Chunqiutang village is considered in this chapter. Households in this village are reasonably representative of the small-holder enterprises in the Red Soils Region and there is, presently, a high level of public interest in using this village as a demonstration village to promote the wider adoption of beef cattle and forages.

Economic Model

The context in which the forage and feeding practices are promoted by the ACIAR project involves immense numbers of resource-constrained households whose welfare is intricately linked to their available farm resources and the wider economy. This context involves complex levels of dependencies between the household and a diverse array of agricultural, craft or industrial activities that can be conducted on or off the farm. There are also many types of interdependencies between farm activities—for example, cropping, forages, livestock—that draw on or supplement available resources, such as animal draught power, feedstuffs, manure and so on. The economic performance of a small-holder enterprise needs, therefore, to be evaluated as an integrated unit. This inherent complexity has been captured within the structure of an Excel® spreadsheet model that includes planted forages and a wide range of other farm and off-farm activities. The structure of the model, which is comprised of three linked worksheets or modules, is presented in Figure 4.2.

FIGURE 4.2
Structure of the Economic Model

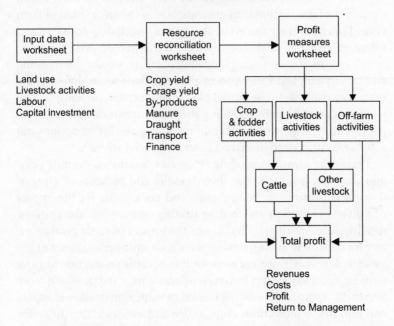

Data relating to the demand for resources by each activity—land, labour, materials, transport services, working capital, feedstuffs, draught and suchlike—or their contribution to farm and household resources—feedstuffs, working capital, manure and so on—are specified in the *Input Data* module. This module also captures data on crop and forage yields, by-products, livestock sales, live-weight gains, reproduction and mortality rates, family structure and labour availability and capital infrastructure, including livestock, plant and equipment, buildings and fences. Limits to purchases of critical resources, such as labour, feedstuffs, manure and draught can also be specified.

The farm and other household activities are linked through resource pools within the *Resource Reconciliation* module, including land, forages and feedstuffs (grain, by-products, straws, oilcake, tops), manure, labour (family and hired), animal draught, and working capital. Final products—for example, crops and sale livestock—and surplus of intermediary produce—such as manure and edible residues—generate revenue; while resource deficits can be filled through purchases from outside the farm, included as expenditures in the calculation of farm costs. The crop, forage and livestock activities—including both final and intermediate farm activities—represent the farm enterprise. Off-farm activities—such as contract ploughing, planting, weeding, harvesting and factory employment—potentially contribute to, or draw on, the resources available to the household for production, consumption or wealth accumulation. By including all of these activities, the model can provide an indication of whether different crop and forage options will actually contribute to or detract from household welfare.

The *Profit Measures* module integrates the input, output, price and cost data from the other two modules and presents an array of summary measures, including gross and net income for the impact of a given production and feeding strategy on the profitability of the small-holder enterprise. To evaluate the impact of cattle production and forage-based feeding strategies on economic performance of the small-holder enterprise, the contribution of cattle production to gross farm income and net farm income is separated from the crops and other livestock. Opportunity costing is used to value non-marketed inputs and outputs to the various crop, forage and livestock activities—for example, draught and manure used for crops—and consumption of crop produce and by-products by livestock. These are treated as transfers of revenue between the producing and consuming activities and give an indication of the real contribution of the activities to the economic welfare of the household. Because these economic transfers do not involve actual exchanges of cash, estimates are also calculated for net cash income, which is the cash that the households would actually receive as a result of employing a given mix of activities.

The model produces an array of profitability measures. The specific profit measures used in the present analysis for the case study farm household include:

1. *Total farm revenue*—gross revenue from the sale of crop, forage, and livestock products, including finished animals, plus the opportunity values of transfers of by-products, residues and other surplus materials between activities.
2. *Total farm costs*—total cash outlays on purchased inputs for farm activities including land lease and produce taxes plus the

opportunity values of transfers of by-products, residues and other surplus materials between activities, depreciation of specialized farm assets, opportunity value of family labour and interest on capital invested in farm assets and livestock. The opportunity costs are based on potential returns from employing the labour and capital assets in a viable non-farm occupation—for example, local labour markets and short-term lending to other farmers.[4]

3. *Return to Management*—Total farm revenue—total farm costs plus income from off-farm activities. This measure is interpreted as the full economic return or profit to the small-holder enterprise after all opportunity costs are accounted for, including the opportunity value of by-products, residues and other surplus materials that are consumed by farm activities; off-farm income and the value of the small-holder family's capital and labour that is utilized for on-farm and off-farm activities.[5]

4. *Net Cash Income*—Total cash revenue—total cash expenses, plus cash income from off-farm activities.

Case Study Household

A case study is presented for a synthetic small-holder farm household assumed to be located in Chunqiutang Village, Hunan Province. Summary data from the survey of 135 households in this village is presented in Table 4.1.

The average household size in this village in 2002 was 4.9 people with a total lease landholding of 5.7 mu (1 mu = 0.07 ha). Approximately two-thirds of the leased land is paddy land and a further 30 per cent is dry land that is suited to growing crops and pasture. Less than one half of the village households—43 per cent actually own cattle, and for these households the average number of cattle is marginally greater than one animal, with a slightly higher proportion being young bulls or steers that are speculatively held for fattening and re-sale. A little over one half of the total household income is comprised of off-farm work. Crop income is almost double that for livestock which is mostly derived from raising pigs and poultry. Income from cattle constitutes less than 10 per cent and 2 per cent of livestock income and total household income. From this data, the model has been calibrated for a representative farm-household whose main characteristics are described in Table 4.2.

TABLE 4.1
Selected Data from a Census Survey of Households in
Chunquitang Village, Dao County, Hunan Province for 2002

No. of households	135
No. of households raising cattle	56
No. of people per household	4.9
Average area of land leased	5.7 mu
– paddy land	3.8 mu
– dry land cultivation	1.6 mu
– ponded land	0.3 mu
Average number of cows per household	0.5 head
Average number of bulls/steers (2 years +) per household	0.8 head
Average number of calves per household	0.1 head
Average number of cattle sold	0.6 head
Average total income per household	9578 yuan
– field crops	2233 yuan
– cattle	134 yuan
– other livestock	1215 yuan
– off-farm income	5640 yuan

Source: MacLeod et al. (2004).

TABLE 4.2
Assumed Characteristics of Synthetic Household, Chungquitang
Village, Dao County, Hunan Province, Used for Scenario Modelling

No. of persons comprising household	4
– working age adults (male and female)	2
– aged adults (60 years +)	1
– children (under 10 years)	1
Number of labour units (adult labour equivalent)	2.75 per annum
Maximum available off-farm labour (adult labour equivalents)	0.8 per annum
Total land leased	7 mu
– paddy land (rice × 2 crops, fallow)	3.8 mu
– dry land cultivation (peanuts × 1 crop, fallow)	3.0 mu
– buildings, paths and the like	0.2 mu
Allocated native grassland	10 mu
Capital invested in farm assets (sheds, pens, tools, barrows etc.)	20,000 yuan
Livestock - cows (aged for draught)	1 head
– feeder steers purchased at 100 kg/live-weight/steer	3 head
– pigs	3 head
– chickens	10 head

Source: MacLeod et al. (2004).

Feed Years

A central task of the ACIAR project was to promote an improved understanding of the year-around feed requirements of cattle given the high level of seasonality in the climate of the Red Soils region and its impact on forage growth and availability. A series of feed year plans have been developed (Nolan et al. 2004) that include planted forages, conserved hay or silage, or other low digestibility feedstuffs, such as rice straw or dry native grass. An example that is applicable to the case study (Hunan Province) is presented in Table 4.3.

TABLE 4.3
**Feed Year Plan for Providing Forage for Small-holder Cattle
Production (3 Head) in Hunan Province** (Value in Tonnes Dry Matter):

Month	Forage	Supplements	Forage Varieties
1.	0.5	0.135	Rice straw, rice bran, cotton seed meal
2.	0.5	0.135	Rice straw, rice bran, cotton seed meal
3.	0.5		ryegrass
4.	0.5		ryegrass
5.	0.5	0.135	Elephant grass hybrid, urea, rice bran, cotton seed meal
6.	0.5	0.135	Elephant grass hybrid, urea, rice bran, cotton seed meal
7.	0.5	0.135	Elephant grass hybrid, urea, rice bran, cotton seed meal
8.	0.5	0.135	Elephant grass hybrid, urea, rice bran, cotton seed meal
9.	0.5	0.135	Elephant grass hybrid, urea, rice bran, cotton seed meal
10.	0.5	0.135	Elephant grass hybrid, urea, rice bran, cotton seed meal
11.	0.5	0.135	Elephant grass hybrid, urea, rice bran, cotton seed meal
12.	0.5	0.135	Elephant grass hybrid, urea, rice bran, cotton seed meal
Total	6.0	1.35	

Scenarios

Four scenarios are presented for the profitability assessment; namely, a baseline case (Scenario 1), which is then modified by increasing the live-weight gain to a level consistent with the average results of the feeding trials (Scenario 2), reducing the average daily live-weight gain of the livestock (Scenario 3), and reducing the sale value of finished stock (Scenario 4)[6].

Scenario 1 (Baseline)

The case study household is comprised of four persons (Table 4.2), which equates to 2.75 labour units that are available for on-farm and off-farm activities. Off-farm labour availability is limited to 0.8 labour units per year. Two rice crops followed by a winter fallow are grown on 3.8 mu of paddy land, and peanuts are grown on 3 mu of cultivable dry land. Cattle are grazed in spring and summer on 10 mu of native grassland available to the household. In addition to three pigs, 10 chicken and an aged cow for draught use, the household purchases three steers at 100 kg per steer average weight. These steers are fed for 200 days on a ration of local forages and crop residues (rice straw and peanut tops). Average daily live weight gain of the steers is 150g/day, which is consistent with the average results for such diets obtained from the feeding trials (Nolan et al. 2004). The finished steers are sold in a local market at 130 kg average live-weight for 5 yuan[7] per kg live-weight. This selling price is consistent with markets for animals of this condition in Dao County in 2005. A summary of other input and output parameter values used in the analysis is presented in Table 4.4.

Scenario 2

In place of the fallow phase, 2.0 mu of ryegrass is grown after two rice crops on the 3.8 mu of paddy land. Hybrid elephant grass is sown on 1 mu of cultivable dry land along with 2 mu of peanuts. The cattle still have grazing access to the 10 mu of native grassland in spring and summer. The 3 steers (100 kg average weight) are fed for 200 days on a mixed ration of forages, residues and supplements consistent with the feed year plan (Table 4.3) at an average daily live-weight gain of 0.6 kg/steer/day and sold locally at 220 kg average live-weight at 6 yuan per kg live-weight. The higher selling price is consistent with the better quality of the finished animals which should attract a premium in local markets.

Scenario 3

Live-weight gain is reduced to 0.5 kg/steer/day, other parameters identical to Scenario 2 (that is, steers are fed for 200 days, selling price

TABLE 4.4
Selected Household Parameter Values for Scenario Modelling

Crops		
	– rice grain yield (tonnes/mu)	
	early crop/late crop	0.4/0.45 tonnes
	– rice grain selling price (yuan/tonne)	
	early crop/late crop	1,200/1,600 yuan
	– peanut grain yield (tonnes/mu)	0.15 tonnes
	– peanut selling price (yuan/tonne)	3,600 yuan
Forages	– astragulus yield (tonnes/dry matter/mu)	0.15 tonnes
	– elephant grass hybrid (tonnes/dry matter/mu)	0.8 tonnes
	– native grasses (tonnes/dry matter/mu)	0.2 tonnes
Fertilizer	– urea application (kg/mu) rice	22 kgs
	– urea price (yuan/kg)	1.7 yuan
	– manure (tonne/mu) all crops	0.2 tonnes
Livestock	– purchase price of feeder steers (live-weight)	6 yuan
	– selling price of feeder steers (live-weight)	6 yuan
	– cattle manure (yuan/tonne/dry weight)	120 yuan
	– draught selling/hiring price (yuan/day)	60 yuan
	– selling price of pigs (yuan/head)	610 yuan
	– selling price of chickens (yuan/head)	12 yuan
Labour	– hire rate (yuan/day)	60 yuan

6 yuan per kilogram). This level of animal performance lies within the lower range of the results from the feeding trials, and is consistent with the limited feeding and husbandry skills of households with little prior experience in raising beef cattle.

Scenario 4
Selling price of the steers is reduced to 5 yuan per kg live-weight, other parameters identical to Scenario 2 (that is, steers are fed for 200 days, live-weight gain 0.6 kg/steer/day). This assumes that the market does not differentiate between different animals on the basis of weight and finish.

Results

The economic results for the four scenarios are summarised in Tables 4.5 to 4.8. The various revenue, cost and profit measures derived by the model are presented for the crop and forage activities and the livestock activities as aggregate measures. Income from off-farm activities is also included to provide an estimate of total household income. The crop and forage activities are combined because they both provide important sources of feedstuffs for the livestock and are generally integrated within the same land area on the farms—for example, sharing draught, labour or inter-cropped. The aggregate livestock measures are also presented separately for the cattle and other livestock activities to provide an indication of the relative contributions of these different activities to farm and household income.

The total revenue and cost measures presented for farm activities (Tables 4.5 to 4.8) include both cash and imputed non-cash items; the latter representing the internal transfers as 'sales' and 'purchases' of the various forages, crop by-products, residues, manure and draught between the crop and forage activities and the livestock activities, depreciation, unpaid household labour and interest on capital. While the resource transfer values offset each other in the calculation of return to management, their identification is necessary to determine the specific contribution of the various activities to the economic performance of the total farm enterprise (Tables 4.5 to 4.8). The effect of including or removing these imputed opportunity values is shown in the differences between the estimates of return to management and net cash income.

Baseline (Scenario 1)

The economic challenges facing small-holder farmers, who may be intending to rear some cattle but have limited skills and experience in growing and feeding forages, are illustrated in the measures of return to management and net cash income (Table 4.5).

The return to management for each of the farm activities, including the cattle rearing activity is negative. While the poor level of animal performance and the cost of labour and capital committed to cattle rearing is the main contributor to this poor farm performance, the crop and forage activities are also failing to cover the cost of the labour and capital committed to them as well. This implies that the small-holder farm household would not be able to indefinitely maintain the investment that it has made in farm production assets, without diverting valuable funds that have been earned through access to the off-farm employment activities. Unless it can increase the productivity of its farm activities—for example, Scenarios 2 to 4—the welfare of the household would be improved by almost 60 per cent if it abandoned them altogether. However, the need for the household to explore these choices may be clouded by the positive contribution of the farm

TABLE 4.5
Revenue, Costs and Profit Measures for Scenario 1

LWG = 0.15 kg/	*Farm activities*		*Livestock component*				
steer/day	*Crops &*	*Total*					*Total*
Price = 6 Y/kg	*Fodder*	*Livestock*	*Cattle*	*Other*	*Total Farm*	*Off-farm*	*Household*
live-weight	*Y Rmb*	*Y Rmb*	*Y Rmb*	*Y Rmb*	*Y Rmb*	*Y Rmb*	*Y Rmb*
Total revenue	6402	3915	1917	1998	10317		
– cash revenue	4818	2655	670	1985	7472		
– resource transfers	1584	1260	1247	13	2844		
Total costs	7067	5130	3104	2024	12197		
– cash costs	1531	1406	1136	269	2937		
– resource transfers	1260	1583	738	845	2844		
– depreciation	500	500	250	250	1000		
– family labour	3276	953	641	312	4229		
– interest on capital	500	687	339	348	1187		
Return to management	–665	–1214	–1188	–27	–1880	3120	1240
Net cash income	3287	1249	–466	1716	4536	3120	7656

activities to net cash income, which is projected to exceed the cash contribution of the off-farm activities by 40 per cent. Nevertheless, the cattle raising activities still generate a negative contribution to cash income, which should either reduce the commitment of the household to cattle raising or spur its' interest in increasing the efficiency of this activity.

Effect of Changing Livestock Performance (Scenario 2)

This option leads to a marked improvement in all of the income measures for the model household (Table 4.6). The return to management from the combined farm activities is positive (cf. Scenario 1), although the return from the cropping and fodder activities remains negative, and for the other livestock activities is small (24 yuan). This implies that these two parts of the farm operation might be profitably reduced in favour of increased commitment to either cattle raising or off-farm activities. The overall economic position for the household remains one in which the contribution of all the farming activities to household

TABLE 4.6
Revenue, Costs and Profit Measures for Scenario 2

	Scenario 2						
LWG = 0.60 kg/	*Farm Activities*		*Livestock component*				
steer/day	*Crops &*	*Total*					*Total*
Price = 6 Y/kg	*Fodder*	*Livestock*	*Cattle*	*Other*	*Total Farm*	*Off-farm*	*Household*
live-weight	*Y Rmb*	*Y Rmb*	*Y Rmb*	*Y Rmb*	*Y Rmb*	*Y Rmb*	*Y Rmb*
Total revenue	6511	6335	4337	1998	12846		
– cash revenue	4818	5056	3075	1980	9874		
– resource transfers	1693	1279	1262	18	2973		
Total costs	7143	5331	3358	1974	12474		
– cash costs	1510	1429	1211	218	2939		
– resource transfers	1279	1694	848	846	2973		
– depreciation	500	500	250	250	1000		
– family labour	3354	953	641	312	4307		
– interest on capital	500	755	408	348	1255		
Return to management	–632	1004	979	24	372	3120	3492
Net cash income	3308	3627	1864	1762	6935	3120	10055

economic welfare remains modest, contributing only 11 per cent to the total net return to management of 3,492 yuan. This economic conclusion, however, might still be masked by the positive levels of net cash income for all of the farming activities. For example, these particularly favour the crop and forage activities by a significant margin, and are more than double the level of off-farm income. Moreover, the net cash income estimates rate the two-livestock activities almost equally and ignore the relatively high imputed value of the provision of draught and manure to the crop and forage activities.

Effect of Changing Livestock Performance (Scenario 3)

Despite being less productive than the previous scenario, this still represents a significant improvement over the baseline animal performance, and is reflected in each of the income measures (Table 4. 7). The cattle raising activity has lifted the overall level of farm performance and the return to management for both livestock activities is positive, indicating that these activities are worth including in the household production mix. The net return to all farm activities is, however,

TABLE 4.7
Revenue, Costs and Profit Measures for Scenario 3

				Scenario 3			
LWG = 0.50 kg/	*Farm Activities*		*Livestock component*				*Total*
steer/day	*Crops &*	*Total*					
Price = 6 Y/kg	*Fodder*	*Livestock*	*Cattle*	*Other*	*Total Farm*	*Off-farm*	*Household*
kive-weight	*Y Rmb*	*Y Rmb*	*Y Rmb*	*Y Rmb*	*Y Rmb*	*Y Rmb*	*Y Rmb*
Total revenue	6512	5885	3887	1998	12397		
– cash revenue	4818	4606	2625	1980	9424		
– resource transfers	1694	1279	1262	18	2973		
Total costs	7168	5295	3313	1981	12463		
– cash costs	1535	1404	1178	225	2939		
– resource transfers	1279	1694	848	846	2973		
– depreciation	500	500	250	250	1000		
– family labour	3354	953	641	312	4307		
– interest on capital	500	744	396	348	1244		
Return to management	–656	590	574	17	–66	3120	3054
Net cash income	3283	3203	1447	1755	6485	3120	9605

negative (–66 yuan) due to the cattle raising activity failing to offset the negative return for crop and fodder activities. Economic performance might be enhanced by reducing the area of land allocated to cropping and forage activities in favour of purchases of crop residues and other forages for the cattle and other livestock activities.

As for the previous Scenario, the net cash income projections may indicate that the crop and fodder activities are an attractive option that contribute almost as much to total household cash income as the off-farm activities. The cash income derived from the other livestock activities also exceeds that for the cattle raising activities by 20 per cent. Small-holders who might make their investment decisions on the basis of cash flow, rather than economic returns, may be attracted towards crop and forage activities and other livestock activities, rather than the higher returning cattle-raising activity.

Effect of Changing Livestock Price (Scenario 4)

As was the case for reducing the production efficiency of the cattle-rasing activity (Scenario 3), reducing the price of the finished animals

(17 per cent to 5 yuan/kg) necessarily reduces the profitability of both this activity and the contribution of the total farm enterprise to the various measures of household income (Table 4.8). Nevertheless, this scenario still represents a marked improvement over the economic performance of the baseline case (Scenario 1, Table 45). However, similar to the outcome for Scenario 3, the combined return to management for the farm activities is negative (−351 yuan), because the reduced return to management for the cattle raising activity is still failing to offset the negative return for crop and fodder activities. Despite this, the cattle raising activity has lifted the overall level of farm performance above the baseline levels (Table 4.5) and the return to management for both livestock activities is positive.

As is the case for each of the previous two scenarios (Table 4.6 and Table 4.7), the contribution of all of the farm activities to net cash income is positive (Table 4.8) and is collectively almost double the level of off-farm income. The skewing of the apparent relative attractiveness of the farm and off-farm activities, based on the net cash income, is also evident in that these activities generate similar levels of net cash income, but have quite divergent economic returns. For example, net cash income favours the cropping and forage activities and other livestock activities over cattle raising, whereas, the economic return to the former is negative and the latter is generating only 4 per cent of the return from the cattle-rasing activity.

Discussion

The rapid expansion of cattle numbers in China over the past 15 years, mainly on small-holder farms with a limited prior history of cattle raising, has been attributed to a favourable perception of the profitability of cattle raising and government encouragement (Longworth et al. 2001). This has been supported by research into growing and feeding high quality forages, such as the ACIAR forage projects in the Red Soils Region. The results of the scenario modelling support both the perception and reality of the economic profitability of cattle-raising activities for small-holder farm households. However, enjoyment of this prospective profitability cannot be guaranteed, but will require more planning and management than simply increasing the number of cattle on a small-holder farm.

TABLE 4.8
Revenue, Costs and Profit Measures for Scenario 4

LWG = 0.60 kg/ steer/day Price = 5 Y/kg live-weight	Farm Activities		Livestock component				Total
	Crops & Fodder Y Rmb	Total Livestock Y Rmb	Cattle Y Rmb	Other Y Rmb	Total farm Y Rmb	Off-Farm Y Rmb	Total Household Y Rmb
Total revenue	6512	5585	3587	1998	12097		
– cash revenue	4818	4306	2325	1980	9124		
– resource transfers	1694	1279	1262	18	2973		
Total costs	7186	5262	3275	1987	12448		
– cash costs	1553	1386	1155	231	2939		
– resource transfers	1279	1694	848	846	2973		
– depreciation	500	500	250	250	1000		
– family labour	3354	953	641	312	4307		
– interest on capital	500	729	381	348	1229		
Return to management	–674	323	312	11	–351	3120	2769
Net cash income	3265	2920	1170	1749	6185	3120	9305

The baseline case (Scenario 1) suggests that including a modest level of cattle raising within the existing farm system without providing adequate feedstuffs is unlikely to be economically viable. This conclusion is consistent with the exploratory analyses of Longworth et al. (2001) whose findings, while not directly focussed on the Red Soils Region, also suggested that the economic returns to beef-raising activities on small-holder farms was at best marginal if not uneconomic. However, that study focussed on conventional feeding practices that typically involved low quality feedstuffs—for example, diets largely comprised of rice straw, other crop residues, and native grasses—with commensurately low levels of assumed animal performance. The ACIAR project is specifically asking whether growing and feeding better quality forages to cattle is economic, answered largely by the results for the remaining three scenarios. The earlier study also did not have access to feeding trials, such as those at Qiyang and Nanchang, which provide critical live-weight performance data for cattle consuming forages under realistic feeding management conditions.

The results for the scenario that is directly based on growing and feeding improved forages (Scenario 2) suggest that the cattle raising activities are a potentially economic option when live-weight gain performance is consistent with the average results for the feeding trials under the prices prevailing during the study period. From a cash-flow perspective, which may attract the interest of resource-poor small-holders, this may also be an attractive one as it offers a substantial increase in the projected level of net cash income. The positive economic outlook for cattle raising is conditional on prevailing market prices for cattle remaining buoyant, and the small-holder households being sufficiently skilled in producing and feeding adequate quality forages to cattle to reap animal performance outcomes similar to those assumed for Scenario 2.

The remaining scenarios (Scenarios 3 and 4) sought to examine the consequences if either of these conditions was challenged, and the general result was similar. For either example, the impact of the adverse change in parameter values was not sufficient to prevent the cattle raising activity from yielding a full economic profit, that is, positive return to management. However, the total mix of farm activities generated an economic loss (negative return to management), because the crop and forage activities yield insufficient income to cover the opportunity costs of labour and capital invested in the farm. This suggests that improving the economic performance of the enterprise may be achieved by seeking improvements in the production and marketing elements of the crop and forage component of the farming system, as well as improving the efficiency of the livestock activities.

Conclusions

Cattle raising activities based on producing and feeding improved forages can potentially increase the economic welfare of small-holder households in the Red Soils Region. However, this is conditional on the efficiency with which such activities are conducted and continued buoyancy in local livestock markets. Under less favourable assumptions of production efficiency and market price, there is scope for economic losses for the farm activity mix, although cattle raising remains profitable. A significant source of economic loss lies in the poor application of valuable livestock services—such as draught and manure—to crop and forage activities. It may be more profitable to divert these resources to off-farm uses—such as hiring out draught or selling manure—or to reduce the level of cropping and divert resources to specialized forage activities or to source feedstuffs from off-farm. A positive economic result is unlikely when cattle raising relies on diets of poor quality feedstuffs (Scenario 1), although positive net cash incomes may continue to generate interest and short-term commitment to this activity.

Interest and official support for beef cattle production activities on small holdings remains high across the Red Soils Region; and a series of extension campaigns supported by demonstration farms in various counties will be implemented to disseminate the results of the ACIAR project feeding trials and feed year plans (CSIRO 2004). A key focus of these campaigns should be to highlight small-holder farm households, the economic imperative of carefully planning forage production and management strategies and to promote appropriate skills in both crop and animal husbandry management. In doing so, the high level of interest and commitment to profitable cattle production in the Red Soils Region can be more appropriately linked.

Acknowledgement

The work described here was funded by the Australian Centre for Agricultural Research as ACIAR project AS/98/35—Ruminant production in the Red Soils Region of southern China and northern Australia.

Notes

1. ACIAR projects PN 89/25, LWR1/93/03, LWR1/96/172.
2. ACIAR project AS2/98/35.
3. The survey of this village was conducted in conjunction with a larger project headed by Professor Chang Songhua, Hunan Agricultural University, to collect baseline production data on beef-raising activities in five Counties in Hunan Province, including Dao County. A census was taken of all households in selected villages in each of the five counties.
4. This definition of total farm costs is more inclusive than standard farm business accounting definitions which typically exclude the imputed value of household labour and interest on capital investment in a step prior to calculating net farm income. However, the present analysis is directly contrasting economic profit (that is, cash and imputed revenue less cash and imputed costs = return to management) with net cash income and the superfluous net farm income calculation is omitted.
5. Given the procedure used to calculate total farm costs (see note 4), this definition is consistent with standard farm business accounting definitions of return to owners'

management (that is, net farm income less imputed values for unpaid householder labour and capital investment). *Note*: there is no opportunity charge for land capital in this return to management calculation, as farmland in the Red Soils Region of China is leased from the State and a lease payment is included in total farm costs.

6. Development costs for establishing the beef enterprise are not specifically accounted for in this study. The four scenarios are all variations on the same structural case which is a representative small-holding that already has the feed, equipment and cattle in place (as reflected in the labour, feed and depreciation charges). The marginal investment between each case is effectively zero as no additional capital expenditure is involved. The actual cost of establishing housing and feed handling facilities for the representative household is likely to be modest, typically involving the use of converted buildings and scavenged materials.

7. one Australian dollar is approximately equivalent to 6.6 Yuan Renminbi in February 2008.

References

Clements, R.J., I.R. Willett, J. Davis and R.A. Fischer. 1997. 'ACIAR's Priorities for Forage Research in some Developing Countries', *Tropical Grasslands* 31: pp. 285–92.

CSIRO. 2004. Final Report—ACIAR Project AS/93/35, *CSIRO Livestock Industries*, Rockhampton.

Longworth, J.W., C.G. Brown and S.A. Waldron. 2001. *Beef in China: Agribusiness Opportunities and Challenges*. Brisbane: University of Queensland Press.

MacLeod, N.D., M. Hu and S. Wen. 2004. Socio-economic Studies. In Report to ACIAR on Project AS/98/35, 'Ruminant Production in the Red Soils Region of Southern China and in Northern Australia', Review 5, *CSIRO Livestock Industries*, Rockhampton. pp. 1–28.

Nolan, J., B. Robertson, Y. He, S. Wen and G. Xie. 2004. 'Cattle Nutrition Research in China'. In Report to ACIAR on Project AS/98/35, 'Ruminant Production in the Red Soils Region of Southern China and in Northern Australia', Review 4, *CSIRO Livestock Industries*, Rockhampton. pp. 1–13.

Sturgeon, J. 1991. 'A Socio-economic Survey in the Red Soils Region.' In Horne, P.M., D.A. MacLeod and J.M. Scott. (Eds) *Forages on Red Soils in China*. Australian Centre for International Agricultural Research, Canberra. pp. 91–102.

Xi, C.F., Q. Xu and J. Zhang. 1990. 'Distribution and Regionalisation of Soils'. In *Soils of China*, Institute of Soil Science, Science Press, Beijing. pp. 20–39.

Xie, K., J. Zhang and P. Horne. 1992. 'Potential and Problems for Forage Development and Animal Production in the Red Soils Region of Southern China'. In Horne, P.M., D.A. MacLeod and J.M. Scott. (Eds) *Forages on Red Soils in China*. Australian Centre for International Agricultural Research, Canberra. pp. 11–14.

Xie, W.M. 1991. 'Developing the Pastoral Industry in the Red Soils Region of Jiangxi Province'. In Horne, P.M., D.A. MacLeod and J.M. Scott. (Eds) *Forages on Red Soils in China*. Australian Centre for International Agricultural Research, Canberra. pp. 117–20.

5

Optimal Land-use with Carbon Payments and Fertilizer Subsidies in Indonesia

RUSSELL M. WISE AND OSCAR J. CACHO

Introduction

Agroforests are often recommended as alternatives to the shifting-cultivation and continuous-cropping systems that are blamed for much of the land degradation in Southeast Asia (Roshetko et al., 2007; Makundi and Sathaye, 2004). But, landholders may not consider tree-based systems as viable alternatives to crops because of high establishment costs, delayed revenues and lack of secure property rights. Recognizing the environmental and social services provided by trees, such as mitigating climate change by sequestering carbon, may assist in overcoming these obstacles.

The conceptual basis of this paper can be illustrated by considering the Production Possibility Frontier (PPF) of a local economy that has a fixed amount of resources to produce bundles of products from two land uses: trees (Y_1) and crops (Y_2) with a given set of inputs and technology (Figure 5.1).

The optimal combination of Y_1 and Y_2 is determined by the price ratio p_1/p_2. If the present value of crop outputs exceeds the present value of tree outputs, the optimal point is likely to be located closer to the

FIGURE 5.1
Pareto Efficient Production Possibilities of an Individual Landholder when
(1) not Receiving Payments for Positive Environmental Externalities
(E_1) and (E_2) when Positive External Effects are Internalized through Carbon-
sequestration Payments (E_2)

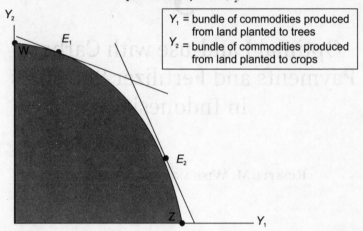

Y_1 = bundle of commodities produced from land planted to trees
Y_2 = bundle of commodities produced from land planted to crops

vertical axis (point E_1); reflecting the current situation in much of the developing world, where continuous cropping is often the preferred land-use option. If the external environmental benefits provided by trees are internalized through direct payments, the price ratio (p_1/p_2) will become steeper and landholders will plant a larger area of their land to trees (point E_2, Figure 5.1).

The Kyoto Protocol (KP) provides the policy context for this analysis, in particular Article 3.3 (Land-use Change and Forestry, LUCF) and Article 12 (Clean Development Mechanism, CDM). These Articles give incentives to developed countries to invest in greenhouse gas mitigation activities, including carbon sinks such as small-scale forestry and agroforestry, in developing countries to help them meet their Kyoto emission limitations at minimum cost. The implications of this are that it becomes possible for landholders to benefit from the resulting technological and financial transfers by claiming credit for sequestered CO_2. Carbon credits[1], however, may only be claimed when sequestered carbon is certified, which requires that project proponents demonstrate a net reduction in emissions compared with the status quo or *baseline*. The effect of the baseline on the eligibility of carbon sequestered by LUCF projects can be significant (Wise and Cacho, 2005a). Here we assume a relatively stable carbon stock, representative of degraded grassland, as the baseline. The problem of the lack of *permanence* of carbon sequestered in biomass and soil is dealt with using the 'ideal' accounting method proposed by Cacho et al. (2003).

In this chapter we develop a meta-model of an agroforestry system and incorporate it into a dynamic programming (DP) algorithm to determine profit-maximizing management strategies in the presence of carbon payments and fertilizer subsidies.

Economic Model

This chapter extends the agroforestry model of Cacho (2001) by including carbon-sequestration payments in addition to the externalities provided by trees on crops. As a starting point, consider a landholder participating in a CDM project and receiving payments for CERs (Certified Emission Reduction). The present value of net revenues (*NPV*) obtained from an area of land A over a project-investment period of T years is:

$$NPV(T,k,x) = (A-k) \cdot \sum_{t=1}^{T} a_t(s_t,k,x_t) \cdot \delta^{-t} + k \cdot \sum_{t=1}^{T} h_t(s_t,k,x_t) \cdot \delta^{-t}$$
$$+ A \cdot \sum_{t=1}^{T} CER_t(s_t,k,x_t) \cdot \delta^{-1} - k \cdot c_E \qquad (1)$$

where s_t represents the state of the land in year t and may be defined by a set of land-quality indicators such as soil depth, soil-carbon content and soil fertility; x is a vector of management decisions such as the timing and frequency of pruning, harvesting and fertilizing activities; k is the area of the farm planted to trees, which remains constant throughout the T years, and $A - k$ is the area planted to crops. The cost of establishing a hectare of trees is c_E and $\delta = (1+r)$ for the discount rate r.

The net annual revenues obtained from the area planted to a single agricultural crop are:

$$a_t = p^a \cdot y_t^a(s_t,k,x_t) - c_t^a \qquad (2)$$

where, y_t^a is crop yield, p^a is the price of the crop and c_t^a is the per-hectare variable costs of preparing the land, sowing seeds, applying fertilizer and harvesting.

The net annual revenues provided by trees are:

$$h_t = p^h \cdot y_t^h (s_t, k, x_t) - c_t^h \tag{3}$$

where, y_t^h is the quantity of tree product harvested in year t, p^h is the price of tree product and c_t^h is the variable costs of harvesting.

The last term in equation (1) is the monetary benefit received for the sale of CERs, which depends on carbon accumulation in tree biomass and soil relative to the baseline (referred to as 'eligible carbon'):

$$CER_t = p^c \cdot (y_t^{bc} (s_t, k, x_t) + y_t^{sc} (s_t, k, x_t)) - cm_t \tag{4}$$

where y_t^{bc} is the change in eligible tree-biomass carbon, y_t^{sc} is the change in eligible soil-carbon stock, p^c is the price of CERs and cm_t is the annual carbon-monitoring cost per hectare.

Equation (1) represents a single rotation and does not include the opportunity cost of keeping trees in the ground. The Faustman model is the standard approach for solving the infinite forestry planning horizon, and it has been extended by authors such as Hartman (1976) to include non-timber benefits (Gutrich and Howarth, 2007). Such models require that the length of each cycle (T), the management variables defined within the vector x, and initial land quality for each cycle S_n remain constant for all cycles $n = 1, 2, \dots N$. These assumptions do not hold when the quality of the land changes over time, possibly resulting in optimal tree areas and rotation lengths changing between cycles. Thus our decision model is:

$$V_n(S_n) = \max_{k_n, x_n, T_n} (NPV_n (S_n, k_n, x_n, T_n) + V_{n+1} (S_{n+1}) \cdot \delta^{-T_n}) \tag{5}$$

subject to:

$$S_{n+1} = S_n + \sum_{t=T_{n-1}+1}^{T_{n-1}+T_n} f_t (s_t, k, x) \tag{6}$$

where, S_n is the quality of the land at the beginning of forestry cycle n, $f_t (\cdot)$ is the annual change in the state variable, and NPV is as defined in equation (1). The problem is solved for an infinite planning horizon of n cycles by backward induction until convergence in $V(S_n)$ is achieved (Kennedy, 1986).

Model Calibration

Agroforests involve growing trees and crops sequentially or simultaneously to improve the productivity and sustainability of the land. Capturing the benefits offered by agroforests necessitates that complementary interactions are maximized and competitive interactions are minimized through management. Agroforestry may involve commercially growing trees with food crops when the trees are young (Otsuka and Place, 2001) or intercopping food crops with nitrogen-fixing trees (Sanchez, 1995). In this study, a rain-fed agroforestry system was investigated in which two maize crops per year were intercropped between Gliricidia (*Gliricidia sepium*) hedgerows over a period of 25 years.[2] The process model SCUAF (Soil Changes under Agriculture, Agroforestry and Forestry) (Young et al., 1998) was used to generate a dataset for meta-modelling. SCUAF was used as it estimates the effects that changes in soil properties (nutrients, soil carbon and soil depth) have on tree and crop productivity based on the management regimes and net-primary-productivity (NPP) rates of the crops and trees. SCUAF has been tested for a range of environments and management conditions by authors such as Nelson et al. (1998) and Vermeulen *et al.* (1993).

Gliricidia was simulated because of its soil-amelioration capabilities and its ability to grow rapidly and produce various commodities such as firewood, fodder or timber. Maize was selected because it is one of the more commonly grown food crops in Indonesia, along with rice, soyabeans and cassava. The parameters selected for this study define a site in a sub-humid climate, with acidic, medium-textured soils of felsic parent material and imperfect drainage. The carbon and nitrogen contents of the system range between 10 and 33 Mg C ha^{-1} and 1.0 and 3.3 Mg N ha^{-1}, respectively—depending on previous land use and degree of degradation. The lower values represent a run-down soil requiring regeneration. Calibration of much of the model was based on data from Nelson et al. (1998) and Grist et al. (1999).

The management parameters varied in this study were area planted to trees (k), fertilizer-application rate (fr), and firewood prune and harvest regime (hr). Total area (A) was set to 1.0, so $0 \leq k \leq 1$ (k also represents a fraction of the area of the small-holding). These values of management parameters were set at the beginning of a simulation and held constant throughout a rotation. A dataset was generated by increasing the area planted to trees (k) at intervals of 0.1, resulting in 11 tree/crop area combinations. Each of these strategies was then replicated under three prune/harvest regimes, resulting in 33 simulated management strategies. In SCUAF, pruning and harvesting intensities are defined as percentages of the annual increment in total tree biomass. In this study, the sum of the prune and harvest intensities was set at 70 per cent of the annual increment in total tree biomass. The remaining 30 per cent of annual biomass increment was not removed from the trees; consequently the carbon contained in trees increased throughout the rotation. Pruned biomass was returned to the soil to decompose and replenish soil carbon and nutrients whereas harvested biomass was removed for sale as firewood. Therefore the soil carbon stock was affected by harvest regime (hr).

Each of the resulting 33 scenarios (11 tree/crop combinations × 3 harvest regimes) was then simulated for four fertilizer application regimes (fr) resulting in a total of 124 treatments. The four fertiliser regimes comprised a mix of nitrogen (N) and phosphorous (P) as follows: (1) $fr = 0$; (2) $fr = 50$ (40 kg N, 10 kg P); (3) $fr = 100$ (80 kg N, 20 kg P); and (4) $fr = 150$ (120 kg N, 30 kg P). These nutrients were added annually to the crop component. According to van Noordwijk et al. (1995), soils in southern Sumatra are often acidic and infertile due to high leaching rates and aluminium toxicity of the subsoil, hence the need for annual fertilizer applications. Nitrogen and potassium deficiencies are probably the most severe constraints on plant productivity making fertilizer application essential.[3] Wise and Cacho (2005a) found that without fertilizer, yields from a Gliricidia-maize agroforest were not sustained beyond the first 10 years of a 25-year simulation period. The biophysical parameter values used in this study were based on Wise and Cacho (2005a; 2005b). The parameter values for the economic model are listed in Table 5.1. Prices are quoted in US dollars using an exchange rate of 10,000 Indonesian Rupiah per US Dollar. A real discount rate of 15 per cent was used to represent the rate of time preference of individual landholders in remote areas of Indonesia (Menz and Magcale-Macandog, 1999).

A simplified econometric production model comprising a set of quadratic equations that interactively mimic soil-carbon changes, tree-biomass accumulation and crop-yield dynamics in response to changes in management was derived based on the dataset generated by SCUAF.

TABLE 5.1
Base-case Parameter Values

Description	Value	Units	Source
Firewood price	4.5	$ Mg⁻¹	a
Price of carbon	15.0	$ Mg⁻¹	d
Price of maize	140.0	$ Mg⁻¹	e
Fertilizer price	0.18	$ kg⁻¹	f
Discount rate	15	%	b
Hedgerow-establishment cost	64.5	$	c
C-monitoring costs	1.0	$ ha⁻¹ yr⁻¹	h
Variable costs for crop	210.0	$ ha⁻¹	c
Price of labour	1.5	$ day⁻¹	g
Maize-harvest labour	5	days Mg⁻¹	c
Prune and harvest labour	3	days Mg⁻¹	c
Labour for weeding	40	days ha⁻¹ yr⁻¹	c
Carbon content of wood	50	%	i

Sources: (a): Wise and Cacho (2005a), (b): Menz and Magcale-Macandog (1999) (c): Nelson et al. (1998) & Grist et al. (1999), (d): Cacho et al. (2003), (e): Katial-Zemany and Alam (2004), (f): USAID, (2003), (g): NWPC, (2005), (h): Wise and Cacho (2005a), (i): Young et al. (1998).

The dataset contained 6,200 observations.[4] The resulting quadratic equations for the state of the soil (s_t) the tree biomass (b_t) and crop yield (y_t^a), respectively are:

$$s_t = \beta_0 + \beta_1 \cdot s_{t-1} + \beta_2 \cdot (s_{t-1})^2 + \beta_3 \cdot s_{t-1} \cdot (1-k)\, \beta_4 \cdot s_{t-1} \cdot fr$$
$$+ \beta_5 \cdot s_{t-1} \cdot hr + \beta_6 \cdot fr + \beta_7 \cdot (1-k) + \beta_8 \cdot (1-k)^2$$
$$+ \beta_9 \cdot (1-k) \cdot hr + \beta_{10} \cdot hr \tag{7}$$

$$b_t = \alpha_0 + \alpha_1 \cdot b_{t-1} + \alpha_2 \cdot (b_{t-1})^2 + \alpha_3 \cdot b_{t-1} \cdot s_t + \alpha_4 \cdot b_{t-1} \cdot k$$
$$+ \alpha_5 \cdot b_{t-1} \cdot hr + \alpha_6 \cdot s_t + \alpha_7 \cdot (s_t)^2 + \alpha_8 \cdot s_t \cdot k + \alpha_9 \cdot s_t \cdot fr$$
$$+ \alpha_{10} \cdot s_t \cdot hr + \alpha_{11} \cdot fr + \alpha_{12} \cdot k + \alpha_{13} \cdot k^2 + \alpha_{14} \cdot hr \tag{8}$$

$$y_t^a = \delta_0 + \delta_1 \cdot s_t + \delta_2 \cdot (s_t)^2 + \delta_3 \cdot s_t \cdot b_t + \delta_4 \cdot s_t \cdot fr + \delta_5 \cdot fr$$
$$+ \delta_6 \cdot \delta_t \cdot fr + \delta_7 \cdot b_t + \delta_8 \cdot (b_t)^2 \tag{9}$$

The explanatory variables in each equation (presented in Table 5.2) are those that fit the simulated treatments best $(P \le 0.05)$. The estimated R^2 and t values reported purely indicate the fit of the quadratic equations to the SCUAF output and are not an indication of the

TABLE 5.2
Base-case Values (Coefficients) for the Dependent Variables of the Quadratic Equations Defining the Biophysical Numerical Model

	Soil Carbon (β)		Tree Biomass (α)		Crop Yield (δ)	
	Coefficient	t-value	Coefficient	t-value	Coefficient	t-value
0	0.7790	(17.18)	−0.8730	(−11.16)	−0.7920	(−7.72)
1	0.9684	(238.65)	0.9910	(628.36)	0.1610	(12.24)
2	0.0004	(4.28)	−0.0048	(−161.99)	−0.0031	(−11.31)
3	0.0062	(8.45)	−0.0005	(−11.59)	−0.0003	(−4.48)
4	−0.00001	(−3.31)	0.2522	(121.85)	0.0001	(5.68)
5	0.00005	(5.25)	−0.0003	(−39.31)	0.0067	(26.72)
6	0.0007	(11.73)	0.0871	(11.55)	−0.0002	(−39.49)
7	−0.6216	(−24.49)	−0.0020	(−12.33)	−0.0370	(−17.56)
8	0.0804	(5.16)	0.0050	(2.88)	0.0010	(23.21)
9	0.0057	(39.99)	0.00002	(4.12)	–	–
10	−0.0066	(−31.12)	−0.0001	(−2.63)	–	–
11	–	–	−0.0004	(−3.49)	–	–
12	–	–	2.7750	(50.42)	–	–
13	–	–	−2.0200	(−40.82)	–	–
14	–	–	0.0020	(4.84)	–	–
R^2		0.99		0.70		0.99

Note: The associated t-values are given as a measure of the significance of each coefficient (a 95 per cent significance requires the t-value be $\geq + 2.08$ or ≤ -2.08).

sampling/measurement errors that is required for statistical inference. This method of approximating a complex, process simulation model with a simple mathematical or econometric model is known as meta-modelling (Kleijnen and Sargen, 2000). Meta-models have been widely used to reduce the time required for full simulation and have been successfully applied to model a variety of environmental problems. Mas et al., (2004), for example, apply meta-modelling techniques to simulate deforestation. Antle and Capalbo (2001) developed such meta-models based on simulated data using the Century model and field-level economic production data, although they refer to such models as 'econometric process' or econometric production simulation' models.

The meta-model, defined by equations (7), (8) and (9), was used to generate values for equations (2), (3) and (4). The crop, wood and carbon yields in these equations were calculated by simple differencing:

$$y_t^{sc} = ((s_t - s_t^0) - (s_{t-1} - s_{t-1}^0)) \tag{10}$$

$$y_t^h = (b_t - b_{t-1}) \cdot hr \qquad (11)$$

$$y_t^{bc} = ((b_t - b_t^0) - (b_{t-1} - b_{t-1}^0)) \cdot \eta \qquad (12)$$

The resulting biophysical and economic outputs were used within the DP model represented by equations (5) and (6).

Results

Optimal decision rules and optimal state transitions were determined by solving the DP model for four carbon-price and fertilizer-price scenarios (Table 5.3), and the effects of tree externalities on the optimal path of the state variable were investigated for the base-case parameters listed in Table 5.1. The low fertiliser price ($p^f = \$0.18\ \text{kg}^{-1}$) represents situations where fertilizers are subsidised (USAID, 2003) and the effect of removing this subsidy is investigated by making $p^f = \$ 0.39\ \text{kg}^{-1}$, which is at the upper range for fertilizer prices as given by van Noordwijk et al., (1995).

Optimal Decision Rules

The optimal tree area ($k*$), cycle length ($T*$), firewood-harvest regime ($hr*$) and fertilizer regime ($fr*$) associated with each of the scenarios in Table 5.3, holding all other variables constant at base-case values, are plotted in Figure 5.2. These plots show the optimal state-contingent decisions. The effect of p^c on optimal management is determined by comparing the solid and dashed curves within each of the eight graphs. The effect of p^f on optimal management is investigated by comparing the graphs between columns 1 and 2 (Figure 5.2).

TABLE 5.3
Four Base-case Carbon- and Fertilizer-price Scenarios Simulated in the Dynamic-programming Model: Each of these is Simulated for a 15 per cent Discount Rate

Scenario	Carbon Price ($ Mg C^{-1})	Fertilizer Price ($ kg^{-1})
1.	15	0.18
2.	0	0.18
3.	15	0.39
4.	0	0.39

FIGURE 5.2
Optimal Management Regimes Obtained by Solving the Dynamic-programming
Model for Four Combinations of Fertilizer and Carbon Prices, at Base-case
Parameter Values

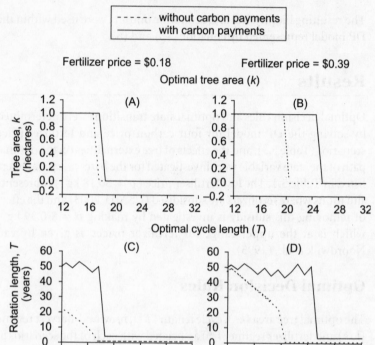

The most significant finding is that it is either optimal to plant only trees or only crops, rather than any combination of the two (Figures 5.2 A & B), which corresponds to points 'w' and 'z' respectively on the PPF in Figure 5.1 and implies a straight-line PPF. Trees are planted when the soil-carbon content is relatively low, because crops are less productive so the opportunity cost of growing trees is low, and to take advantage of the trees' ability to restore the soil through nitrogen-fixation and residue additions (Figures 5.2 E & F). The higher the p^c and p^f the greater the stock of soil carbon required before the optimal solution switches from trees to crops because the opportunity cost of switching to crops is higher. Fertilizer is not used when growing trees because Gliricidia is nitrogen-fixing.

In the absence of carbon payments, and with a low fertilizer price (left panel in Figure 5.2), it is optimal to plant the entire plot to trees s_t at values less than about 17.5 Mg C ha^{-1} (Figure 5.2 A) for rotations of between 7 and 22 years (Figure 5.2 C), to return 80 per cent of pruned biomass to the soil as residues (Figure 5.2 G) and to not apply fertilizer (Figure 5.2 E). It is optimal to do this because the soil is not productive enough to produce acceptable maize yields, even when fertilizer is used. However, at values of greater than 17.5 Mg C ha^{-1}, it is optimal to grow crops continuously and to apply 150 kg ha^{-1} of fertilizer because larger profits are made and maize yields can be sustained.

With unsubsidized fertilizer (\$0.39) and without carbon payments (right panel in Figure 5.2) similar optimal-decision rules are observed but the lines shift to the right and s_t must now exceed 20.5 Mg C ha^{-1} to make crops the optimal land use. At a higher p^f the optimal cycle length increases to between 22 and 48 years, depending on the initial amount of carbon in the soil (Figure 5.2 D). Longer tree cycles are optimal because more time and tree biomass are required to increase s_t to 20.5 Mg C ha^{-1} than to 17.5 Mg C ha^{-1} as required at a low p^f. Further, the higher p^f makes the opportunity cost of planting trees lower.

Carbon payments provide incentives to keep trees for longer and at higher soil-carbon levels (compare solid lines with dashed lines in Figure 5.2). It is now optimal to grow trees for s_t values up to 18.5 Mg C ha^{-1} for the low p^f and up to 25.5 Mg C ha^{-1} for the high p^f and to increase tree-cycle length to between 41 and 50 years depending on p^f. The critical value of s_t at which it becomes optimal to switch from trees to crops increases in the presence of carbon payments because a more productive soil is needed to make crops more profitable than trees.

Optimal-State Paths

The trajectories of the state variable (s_t) that result from applying the optimal-decision rules over a period of 150 years are plotted in Figure 5.3. If the initial soil quality is relatively good ($S_0 = 33$ Mg C ha^{-1}), it is optimal to exploit the system with continuous cropping and fertilizer application while reducing soil carbon for 57 years until it reaches an equilibrium value of 27.8 Mg C ha^{-1}. When the initial soil quality is relatively poor ($S_0 = 12$ Mg C ha^{-1}), it is optimal to build up soil carbon to a plateau (17.8, 22.8, or 28.1 Mg C ha^{-1} depending on p^f and p^c) by growing trees and returning pruned biomass to the system as residues, and then switching to crops plus fertilizer.

FIGURE 5.3
Optimal State Paths Associated with the Optimal Management Decisions Obtained by Solving the Dynamic-programming Model for Four Combinations of Fertilizer and Carbon Prices and Two Levels of Initial Soil Carbon, at Base-case Parameter Values

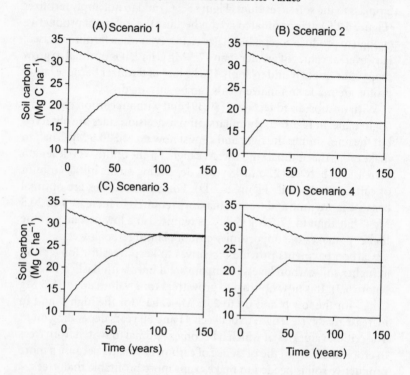

With $S_0 = 33$ Mg C ha^{-1}, the presence of carbon payments and/or the removal of fertilizer subsidies has no effect on the optimal soil-carbon path; it is optimal to plant crops and not to participate in the carbon market. When the system is relatively degraded ($S_0 = 12$ Mg C ha^{-1}), it is optimal to grow trees to replenish soil-carbon stocks and to participate in the carbon market. When s_t reaches its target equilibrium, crops are grown because the opportunity cost of growing trees has increased as a result of the higher soil carbon. This means that the initial state of the soil (S_0) as well as prices influence the optimal level of soil carbon at equilibrium. Only when a high p^f is combined with carbon payments is it optimal to build soil carbon to a single equilibrium level (Figure 5.3 C). The decisions that cause the s_t trajectories depicted in Figure 5.3 may be determined from Figure 5.2, where the optimal decision rules, associated with all states of the soil, are plotted.

Finally, it is informative to investigate the trajectories of the total eligible-carbon stock associated with the optimal-decision rules, as this reflects the cumulative stream of annual carbon payments (Figure 5.4). The trajectories of the eligible-carbon stock emphasise the positive relationship between p^c and p^f on the quantity of CERs associated with each optimal management regime.[5]

FIGURE 5.4
The Trajectory of the Eligible–Carbon Stock Associated with the
Optimal Management Regimes for the Different Prices of
Carbon and Fertilizer for a Poor Quality Soil

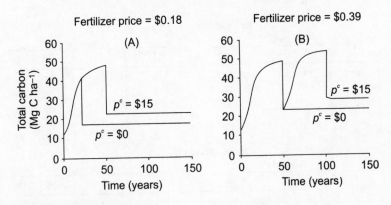

Sensitivity Analysis

The optimal decisions (for the first five cycles only) for both low (columns 2 to 5 in Table 5.4) and high soil carbon values (columns 8 to 11 in Table 5.4) that lead to the state paths are shown in Figures 5.3 and 5.4 respectively. The sensitivity of these optimal decisions to an increase in the price of fertilizer due to higher fuel prices was investigated by running the model for two further carbon-price and fertilizer-price scenarios (tabulated as scenarios 5 and 6 in Table 5.4). The fertilizer price was assumed to increase to US\$ 0.6, and all other prices remained unchanged. Scenario 5 represents the optimal decisions at a higher fertilizer price and with $p^c =$ US\$ 15. Scenario 6 represents the higher fertilizer price with $p^c =$ US\$ 0.

TABLE 5.4
Optimal Decisions Over Five Cycles for Base-case Fertilizer- and Carbon-price Scenarios, and Two High Fertilizer-price Scenarios (Scenarios 5 and 6), at a Low and High Initial Soil–Carbon Level, and a Discount Rate of 15 per cent

| Cycle | Low Initial Soil Carbon ($s_0 = 12$ Mg C ha^{-1}) | | | | | | High Initial Soil Carbon ($s_0 = 33$ Mg C ha^{-1}) | | | | | |

Optimal tree area (k*)

Cycle	Scenario						Scenario					
	1	2	3	4	5	6	1	2	3	4	5	6
1	1	1	1	1	1	1	0	0	0	0	1	1
2	0	0	1	0	1	1	0	0	0	0	1	1
3	0	0	0	0	1	1	0	0	0	0	1	1
4	0	0	0	0	1	1	0	0	0	0	1	1
5	0	0	0	0	1	1	0	0	0	0	1	1

Optimal rotation (T*, yrs)

Cycle	Scenario						Scenario					
	1	2	3	4	5	6	1	2	3	4	5	6
1	49	21	49	49	49	49	1	1	1	1	31	49
2	1	1	50	1	50	47	1	1	1	1	36	50
3	1	1	1	1	37	50	1	1	1	1	42	50
4	1	1	1	1	43	45	1	1	1	1	31	49
5	1	1	1	1	42	46	1	1	1	1	43	50

Optimal harvest (hr*, %)

Cycle	Scenario						Scenario					
	1	2	3	4	5	6	1	2	3	4	5	6
1	20	20	20	20	20	40	0	0	0	0	40	60
2	0	0	20	0	20	80	0	0	0	0	40	60
3	0	0	0	0	20	20	0	0	0	0	20	20
4	0	0	0	0	40	100	0	0	0	0	40	20
5	0	0	0	0	20	20	0	0	0	0	40	80

Optimal fertilizer (fr*, kg)

Cycle	Scenario						Scenario					
	1	2	3	4	5	6	1	2	3	4	5	6
1	0	0	0	0	0	0	150	150	150	150	0	0
2	150	150	0	150	0	0	150	150	150	150	0	0
3	150	150	150	150	0	0	150	150	150	150	0	0
4	150	150	150	150	0	0	150	150	150	150	0	0
5	150	150	150	150	0	0	150	150	150	150	0	0

The most noticeable change, for both initial soil carbon levels, is that it is never optimal to plant crops but to grow trees only and use no fertilizer. When growing trees is the landholder's only option, the intensity with which the trees are harvested becomes extremely important as the landholder must now balance the tradeoffs between harvesting the wood for firewood (and depleting the soil of carbon) or not harvesting and returning the biomass to the soil to increase the carbon stock of the system (and forgoing revenue from firewood sales). This decision depends on whether the opportunity to receive carbon payments exists or not. When receiving carbon payments (scenario 5), the optimal decision path involves harvesting an average of only 20 per cent (when s_0 is low) or 40 per cent (when s_0 is high) of the annual increment in aboveground biomass.

When not receiving carbon payments (scenario 6), incentives exist to exploit the system and the optimal harvest regime increases to an average of about 50 per cent of the annual increment in aboveground biomass, irrespective of initial soil carbon stock. The tree-rotation lengths are not particularly responsive to either initial soil-carbon stock or to whether carbon payments are received or not.

Discussion and Conclusions

Our results indicate that the optimal-decision path through time depends on the initial carbon content of the soil (s_0). If the land has relatively high soil-carbon content, it is optimal to only grow crops and to apply fertilizers. The crops initially deplete the soil of carbon until a 'target' steady state is reached where it is then maintained over time. In this case, carbon payments have no effect on the optimal management of the system; but they do decrease profitability because landholders are required to pay for the carbon lost from the soil. Consequently, based on the assumptions of this chapter, incentives do not exist for landholders to participate in the carbon market when soil quality is good. This is especially true when fertilizer is subsidized, as this increases crop profitability.

If the initial soil quality is relatively poor the results are quite different. Optimal management involves planting the entire area with trees for cycles lasting 20 to 100 years and returning 80 per cent of pruned biomass to the soil to replenish soil nutrients. This increases the soil-carbon stock and the productivity of the system. Once the trees have built up the soil-carbon stock to a target steady state, it becomes optimal to switch to only crops and to use fertilizer to help maintain the soil-carbon level. The optimal number of tree rotations and their optimal length depend on carbon and fertilizer prices. Payments for carbon make it optimal to lengthen the tree cycle and, if combined with a high fertilizer price, it becomes optimal to plant a second tree rotation. It is always optimal to participate in the carbon market when growing trees.

An important finding in this analysis is that it is generally not optimal to build poor quality soils (low soil-carbon content) up to the same target steady state as that reached for good quality soils. The target steady state to which the carbon content of poor quality soils is raised depends on the prices of carbon and the fertilizer. Only when carbon and fertilizer prices are high is it optimal to build a low s_0 up to the same target steady state as that reached for soils with high s_0.

The currently high fuel prices may affect not only the cost of producing fertilizers, but also the cost of transportation to remote areas with poor roads. It was found that at a fertilizer price of $ 0.6/kg it is never optimal to plant crops, but it is optimal to grow trees only and to use no fertilizer, irrespective of the initial soil carbon stock. When growing only trees, the intensity with which the trees are harvested becomes extremely important as the landholder must balance the trade-offs between harvesting the wood for firewood (and depleting the soil of carbon), or not harvesting and returning the biomass to the soil to increase the carbon stock of the system (and forgoing the revenue from firewood sales). This decision depends on whether the opportunity to receive carbon payments exists or not.

Finally, this study has identified issues requiring further investigation. Firstly, under certain economic and biophysical conditions it was optimal to grow only trees for periods between 20 and 100 years. Such a commitment has implications for landholder food security and traditional farming of food crops and it may be unlikely that farmers will adopt such practices. This might be overcome by adopting a landscape approach to land management whereby trees are planted to restore degraded areas while crops are planted in the better land. Secondly, property rights associated with trees and tree products often do not exist or are poorly defined in developing countries, which is likely to make the long-term adoption of trees unlikely unless the appropriate institutional arrangements are in place. Thirdly, the risks of growing trees (for example, fires and illegal logging) have not been included in the model but may alter the decision rules found to be optimal. Fourthly, the implications of payments for the emission-reduction generated, when the firewood harvested is used to substitute for fossil fuels, needs to be investigated. Lastly, the optimal decision rules and state paths identified for the assumptions in this study imply that the PPF of the simulated agroforestry system is a straight line because corner solutions were always obtained. The implications of assuming a more conventional PPF (for example, a system with stronger complementarities between trees, soils and crops) can be investigated by modifying the parameters of the meta-model.

Notes

1. The proposed medium of exchange of C credits under the CDM is the Certified Emission Reduction (CER).
2. Many cropping patterns exist in dryland areas of Indonesia, including sequential plantings of maize (Fagi, 1992) and relay cropping of maize, soyabean (*Glycine max*) and velvet bean (*mucuna pruriens*) (Sitompul et al., 1992).
3. Nelson et al. (1999) and Fagi (1992) recommend between 120 and 248 kg urea ha^{-1} yr^{-1} and between 93 and 98 kg of triple super phosphate (TSP) be added when growing two maize crops ha^{-1} yr^{-1}.
4. The product of 124 management regimes, 25 years and two initial states of soil quality.
5. The equivalent trajectories when s_0 is high are not plotted as they are identical to those presented in Figure 5.4.

References

Antle, J. and S. Capalbo. 2001. 'Econometric-process Models for Integrated Assessment of Agricultural Production Systems', *American Journal of Agricultural Economics*, 83: pp. 389–401.

Cacho, O.J. 2001. 'An Analysis of Externalities in Agroforestry Systems in the Presence of Land Degradation', *Ecological Economics*, 39 (1): pp. 131–43.

Cacho, O.J., R.L. Hean and R.M. Wise. 2003. 'Carbon-accounting Methods and Reforestation Incentives', *The Australian Journal of Agricultural and Resource Economics* 47 (2): pp. 153–79.

Fagi, A.M. 1992. *Upland Crop Production in Indonesia with Regard to Fertilizer Utilization*, Sukamandi Research Institute for Food Crops. Agency for Agriculture Research and Development, Indonesia.

Grist, P., K. Menz and R. Nelson. 1999. 'Gliricidia as Improved Fallow', in K. Menz, D. Magcale-Macandog and I. Wayan Rusastra (ed.), *Improving Smallholder Farming Systems in Imperata Areas of Southeast Asia: Alternatives to Shifting Cultivation*, ACIAR Monograph No. 52: pp. 133–47.

Gutrich, J. and R.B. Howarth. 2007. 'Carbon Sequestration and the Optimal Management of New Hampshire Timber Stands, *Ecological Economics*, 62: 441–50.

Hartman, R. 1976. 'The Harvesting Decision when a Standing Forest has Value', *Economic Inquiry* 14 (1): pp. 52–80.

Katial-Zemany, A. and N.S. Alam. 2004, Indonesia Grain and Feed Annual GAIN (Global Agriculture Information Network) Report, 2004, ID4011, Jakarta, USDA Foreign Agricultural Service.

Kennedy, J.O.S. 1986. *Dynamic Programming: Applications to Agriculture and Natural Resources*, London Elsevier Applied Science Publishers.

Kleijnen, J. and R. Sargen. 2000. 'A Methodology for Fitting and Alidating Metamodels in Simulation', *European Journal of Operational Research*, 120: pp. 14–29.

Makundi, W. and J. Sathaye. 2004. 'GHG Mitigation Potential and Cost in Tropical Forestry-relative Role for Agroforestry', *Environment, Development and Sustainability*, 6 (1–2): pp. 235–60.

Mas, J.F., H. Puig, J.L. Palacio and A. Sosa-Lopez. 2004. 'Modelling Deforestation using GIS and Artificial Neural Networks', *Environmental Modelling and Software*, 19 (5): pp. 461–71.

Menz, K. and D. Magcale-Macandog. 1999. 'Introduction', in K. Menz, D. Magcale-Macandog and I. Wayan Rusastra (ed.), *Improving Smallholder Farming Systems in Imperata Areas of Southeast Asia: Alternatives to Shifting Cultivation*, ACIAR Monograph No. 52: pp. 1–10.

Nelson, R., R.A. Cramb, K.M. Menz and M.A. Mamicpic. 1998. 'Cost-benefit Analysis of Alternative Forms of Hedgerow Intercropping in the Phillipine Uplands', *Agroforestry Systems*, 39 (3): pp. 241–62.

———. 1999. 'Gliricidia, Napier and Natural Vegetation Hedgerows', in K. Menz, D. Magcale-Macandog, and I. Wayan Rusastra (ed.), *Improving Small-holder Farming Systems in Imperata Areas of Southeast Asia: Alternatives to Shifting Cultivation*, ACIAR Monograph No. 52: pp. 95–112.

NWPC. *Comparative Wages in Selected Asian Countries*. 2005. National Wages and Productivity Commission (NWPC), Republic of the Philippines, Department of Labour and Employment, Available from http://www.nwpc.dole.gov.ph/pages/statistics/stat_comparative.html.

Otsuka, K. and F. Place. 2001. *Land Tenure and Natural Resource Management: A Comparative Study of Agrarian Communities in Asia and Africa*, Baltimore: Johns Hopkins University Press.

Roshetko, J.M., R.D. Lasco and M.S. de Los Angeles. 'Smallholder Agroforestry Systems for Carbon Storage', *Mitigation and Adaptation Strategies for Global Change* 12 (2): 219–42.

Sanchez, P.A. 1995. 'Science in Agroforestry', *Agroforestry Systems.* 30 (1–2): pp. 1–55.

Sitompul, S.M., M.S. Syekhfani and J. van der Heide. 1992. 'Yield of Maize and Soyabean in a Hedgerow Intercropping System', *Agrivita*, 15 (1): pp. 69–75.

USAID, Indonesian Food Policy Program. 2003. Policy Brief No. 39. Fertiliser Prices and Fertiliser Imports, Food Policy Advisory Team, USAID, DAI.

van Noordwijk, M., S.M. Sitompul, K. Hairiah, E. Listyarini and M.S. Syekhfani. 1995. 'Nitrogen Supply from Rotational or Spatially Zoned Inclusion of Leguminosae for Sustainable Maize Production on an Acid Soil in Indonesia', in R.A. Date (ed.), *Plant Soil Interactions at Low pH*, Kluwer Academic Publishers. pp. 779–84.

Vermeulen, S.J., P. Woomer, B.M. Campbell., W. Kamukondiwa, M.J. Swift, P.G. H. Frost, C. Chivaura, H.K. Murwira, F. Mutambanengwe and P. Nyathi. 1993. 'Use of the SCUAF Model to Simulate Natural Miombo Woodland and Maize Monoculture Ecosystems in Zimbabwe', *Agroforestry Systems*, 22, pp. 259–71.

Wise, R.M. and O. J. Cacho. 2005a. 'A Bio-economic Analysis of Carbon Sequestration in Farm Forestry: a Simulation Study of Gliricidia Sepium', *Agroforestry Systems* 64: pp. 237–50.

———. 2005b. 'Tree-Crop Interactions and Their Environmental and Economic Implications in the Presence of Carbon-Sequestration Payments', *Environmental Modelling and Software*, 20: pp. 1139–48.

Young, A., K. Menz, P. Muraya and C. Smith. 1998. 'SCUAF Version 4: A Model to Estimate Soil Changes under Agriculture, Agroforestry and Forestry', *Technical Reports Series*. 41. Canberra, Australia, Australian Centre for International Research (ACIAR).

6

Deterioration of Tank Irrigation Systems and Poverty in India

KEI KAJISA

Introduction

International attention to water scarcity in developing countries has been increasing (as can be seen from the establishment of the series of World Water Forums), and with this increasing attention has come the realization that efficient water management is crucial for sustainable development. In the agricultural sectors of developing Asian countries, a major change, in recent times, in water management systems is the rapid spread of pumps and wells (modern irrigation systems) and the decline of traditional irrigation systems such as tanks in Tamil Nadu, India. Facing this transition with increasing water scarcity, people are raising questions on whether water is being used efficiently under the modern irrigation systems and how the decline in the traditional systems affects the poor farmers who are usually the last group of people to adopt the modern systems. This chapter aims to investigate these questions using a case in Tamil Nadu.

Tank irrigation systems collectively operated and managed by informal local bodies have been a dominant source of irrigation in southern India since time immemorial (Palanisami 2000). However, in the last two decades, a massive diffusion of private wells together with pumps has occurred throughout India due to a sharp decline in investment and operation costs; Tamil Nadu has been no exception (Kajisa et al., 2007). In Tamil Nadu, the per cent share of agricultural area under well irrigation increased from 26 per cent in the 1960s to 42 per cent in the 1990s, while the share under tanks declined from 38 per cent to 22 per cent over the same period (Fertilizer Association of India, various issues). The shift in relative share reflects not only a preference for private wells but also the actual replacement of tank systems by well systems. This replacement process has been associated with significant increases in the average yield of rice, a staple crop in the area, and in the average income level of farmers.

Despite such positive effects on the average, there is concern that the replacement of tanks by private wells is associated with increased poverty. Access to water from tanks is available to all the farmers in the system command area, in principle. Access to irrigation water from private wells, however, is limited to owners, and to those who can buy from the owners. Private wells provide freedom in irrigation water control and thus those who have access can increase their yield and income. Since the tank is an indivisible technology, when farmers with access to private wells exit from the collective management of their tank system—out of disinterest or loss of incentive—it becomes difficult for the remaining smaller number of users to provide a sufficient level of maintenance work (Kajisa et al., 2007). When the decline in collective management happens, farmers who are dependent on tanks suffer, while farmers who have recourse to private wells can still achieve high levels of income and crop yield. In this way, the farmers without access to private wells suffer negative effect created by other farmers' exit from the collective management, leading to increased poverty among the without-access farmers.

The story does not end here; a problem also arises among well users. Since the groundwater is a typical example of common pool resources under open access, each individual user, in his/her efforts at maximizing profit, does not take account of the existence of a negative externality he/she imposes on other users, resulting in the exploitation of groundwater beyond a social optimum level. Therefore, eventually, the well users become unable to earn as much profit as they used to on rice.

Based on the argument above, we advance two hypotheses. First, once the decline in collective management occurs, the rice yield and income of the non-well users alone will decrease, resulting in increased poverty among them. The second hypothesis to be tested is that the dissemination of private wells leads to the overexploitation of groundwater, resulting in no significant increase in profit on rice among the well users. As these hypotheses were found to hold, we argue that the rural areas in Tamil Nadu suffer from double tragedies—not only an increased poverty among non-well users but also a reduced profit from rice cultivation among well users.

The next section describes the study site and data collection. Sections three and four test our first and second hypotheses respectively. Section five summarizes the results and presents policy implications.

Study Site and Data Collection

This study is based on our survey of 79 tank-irrigated villages randomly selected from four contiguous districts (Madurai, Ramnad, Virudunagar and Sivaganga) in the southern part of the state of Tamil Nadu, India. In these districts, rice is the dominant crop, irrigated mainly by tanks supplemented by wells. Other crops such as sorghum, millet, groundnut, cotton, chilli and sugarcane are cultivated with or without irrigation. In each village, we conducted a group interview to collect information on the management of tank irrigation as well as on village characteristics. Where a village used several tanks, we identified the most important one through the group interview and collected information on that particular tank. We also interviewed 450 rice-farming households, randomly selected with a sample of five or six households from each village. Due to the focus of this study, we excluded non-farmers from the sample. The field survey was conducted by Tamil Nadu Agricultural University in 1999. Normal rainfall was recorded at the study sites in the survey year.

The Impact of the Decline in Collective Management on Poverty

Measuring Collective Management

To evaluate the impact of the decline in collective management, the first step we have to take is to measure the overall status of the collective actions devoted to tank management in the survey year as compared with the past. There are two approaches: (1) evaluating the condition of irrigation systems as a resulting indicator of effective collective management; and (2) evaluating the degree of cooperation within the collective management.[1] The former approach is not appropriate for our case due to difficulties in isolating the current status of irrigation from the influence of exogenous environmental conditions and from accumulated past successes or failures in collective management. Therefore, we use the latter approach, although this approach also presents difficulties. Past studies generate a dichotomous variable because of the difficulty in objectively ranking the degree of cooperation. We could generate a dummy for each of the three tasks of collective management: de-silting, channel cleaning and the arrangement of water distribution. However, evaluating each separately will not necessarily provide useful information regarding overall status because the activities may be mutually substitutable, so that the lack of one activity will not necessarily mean an overall decline (Fujiie et al., 2005). Moreover, the lack of de-silting and channel cleaning does not necessarily mean that management declined in the survey year, because those activities are carried out according to need.[2] To ensure that the second approach produces appropriate measurements, we must be able to measure the overall status in the survey year.

The approach we take is to use the dichotomous response of key village informants to the question of whether the informal water users' organization (WUO) is active or inactive in the survey year; the dummy takes the value one if inactive. We consider this dummy variable to be an appropriate proxy for measuring tank management activity because firstly, it evaluates the overall status of the collective management, and secondly, it evaluates the status in the survey year. Although this variable is somewhat arbitrary, it is the best available proxy that we could think of, having high correlation with other proxy variables.[3]

Using this dummy as our measure, we classified the villages into those with and those without active collective management, and we found that the number of 'inactive villages', (that is, without active collective management) is 31 (39 per cent) and the number of 'active villages' is 48 (61 per cent).

Binary Comparison

In order to develop an understanding of the impact of the decline in collective management, we compare yield, per capita monthly income, per capita monthly consumption expenditure and a subjective poverty assessment among 171 households in the 31 inactive villages and 279 households in the 48 active villages.[4] These data are displayed in Table 6.1. The difference in rice yields is clear: row (1) shows that

TABLE 6.1

Comparison of Rice Yield, Income, Consumption Value and Subjective Poverty Assessment of Sample Households between Collective Management Inactive Villages and Active Villages

	(A) Households in Inactive Villages (Village: n = 31) (HH: n = 171)	(B) Households in Active Villages (Village: n = 48) (HH: n = 279)	Difference (A)–(B) ($\|t\text{-ratio}\|$)
1. Rice yield (kg/hectare)			
Mean	3499	3786	−287 (2.59)**
Std. Dev.	1275	1049	
2. Per capita monthly income (Rs/person)[a]			
Mean	332	395	−63 (1.77)*
Std. Dev.	370	363	
Head count ratio of poverty (P_0)[b]	58.5	48.3	10.2
Poverty gap (P_1)[b]	33.7	21.6	12.1
3. Per capita monthly consumption expenditure (Rs/person)[a]			
Mean	308	336	−28 (1.46)
Std. Dev.	194	202	
Head count ratio of poverty (P_0)[b]	52.0	45.5	6.5
Poverty gap (P_1)[b]	19.8	14.2	5.6
4. Subjective poverty assessment Percentage of villages assessing that the current condition of poverty is serious (%)	61	29	

Notes: [a] The value is converted into per capita base using the adult equivalent number of present household members. See note 4 for details of the conversion method.

[b] International poverty line of US$ 1 per day adjusted for purchasing power parity is used.

* significant at 10% level; ** significant at 1% level.

rice yield in kg per hectare is lower in the inactive villages than in the active villages and the difference is statistically significant. The lower average yield leads to lower average income and consumption expenditure as shown in rows (2) and (3).[5] The comparison of poverty indexes shows that not only the incidence of poverty but also the poverty gap is higher in the inactive villages.[6] These poverty indexes are consistent with the villagers' subjective assessment of their poverty conditions presented in row (4); a larger percentage of villagers in the inactive villages judged that they are in serious poverty when compared to villagers in the active villages. These results imply the existence of impact of the decline on increased poverty.

Differential Impact of the Decline in Collective Management

Our first hypothesis claims that the decline in collective management reduces rice yield and income/consumption only among farmers who have no access to private wells. The results in Table 6.1 imply the validity of this hypothesis but do not directly examine this hypothesis because the difference in the accessibility to private wells is not incorporated in the analysis. In order to examine the differential impact of the decline between the farmers with access and those without access, we classify farmers into four groups: (1) farmers with access to wells at a collective management active village, (2) those with access at an inactive village, (3) those with no-access at an active village, and (4) those with no-access at an inactive village. If our first hypothesis holds, we will find no statistical difference in yield and income/consumption between Groups 1 and 2 because both of them can access wells, while we will find that yield and income/consumption of Group 4 are significantly lower than those of Group 3.

In order to classify farmers into these four groups, we need to distinguish farmers with access and those with no access. In our analysis, non-well-owners who did not buy water from well owners are classified as the no-access group, while the owners and the non-owners with water transaction records are classified as the access group.[7] Out of 450 sample farmers, 86 are classified as the first group, 40 as the second, 193 as the third and 131 constitute the fourth group.

Table 6.2 shows the results for rice-yield comparison between the four groups. The mean values are reported in the second column. Columns three to five report the differences of mean values and their statistical significance. For example, the difference from Group 1 to Group 2 is 288 (that is, 4,392 − 4,104 = 288) and the absolute value of t-ratio for the test of no difference is 1.54. Since this difference is not statistically significant at any conventional level, the result indicates that even if the collective management declines, farmers can

TABLE 6.2
Comparison of Rice Yield between Different Irrigation Statuses

(A) Irrigation Status	Mean Value (St. Dev.)	*Difference of the Means between Irrigation Status (A) and Irrigation Status below (B)*		
		Mean of (A) – Mean of (B) (|t-ratio|)		
		(B)		
		(1)	*(2)*	*(3)*
		Well Access Active Collective Management	*Well Access Inactive Collective Management*	*No Well Access Active Collective Management*
1. Well access Active collective management	4104 (976)			
2. Well access Inactive collective management	4392 (964)	288 (1.54)		
3. No well access Active collective management	3644 (1051)	–460 (3.44)*	–748 (4.15)*	
4. No well access Inactive collective management	3225 (1235)	–879 (5.54)*	–1167 (5.48)*	–419 (3.27)*

Note: * significant at 1% level.

achieve the same high yield as long as they can access private wells. The row of Group 3 shows that the yield of Group 3 is lower than that of Group 1 by 460 kg and much lower than that of Group 2 by 748 kg. These results indicate that even if farmers can get irrigation water from properly maintained tanks, their yield cannot be as high as those with access to private wells because well irrigation systems indicate superior technology in terms of water control. The yield of Group 4, the most water scarce group, is significantly lower than any of the other groups. Among the findings above, the most important one is that the decline in collective management does not reduce yield if farmers can access private wells, while it significantly reduces yield if they cannot do so.

Tables 6.3 and 6.4 compare income and consumption expenditure respectively. The statistical results are essentially the same as those of

TABLE 6.3
Comparison of Income Per Capita between Different Irrigation Statuses

		Difference of the Means between Irrigation Status (A) and Irrigation Status below (B)		
		Mean of (A) – Mean of (B) (|t-ratio|)		
		(B)		
(A) Irrigation Status	Mean Value (St. Dev.)	(1) Well Access Active Collective Management	(2) Well Access Inactive Collective Management	(3) No Well Access Active Collective Management
1. Well access Active Collective Management	588 (464)			
2. Well access Inactive collective management	560 (416)	−28 (0.32)		
3. No well access Active collective management	308 (266)	−280 (6.35)*	−252 (4.88)*	
4. No well access Inactive collective management	262 (325)	−326 (6.08)*	−298 (4.74)*	−46 (1.41)+

Notes: * significant at 1% level; + significant at 10% level (one-sided test).

Table 6.2. One exception is that the comparison of income between Group 4 and Group 3 (Table 6.3) shows a less statistically significant result (the difference of −46 with t-ratio of 1.41). Nevertheless it can be still significant at one-sided test. Hence, based on the results in Tables from 6.2 to 6.4, we conclude that our Hypothesis 1 holds.

The Diffusion of Private Wells and Rice Profit among Well Users

To test our second hypothesis, we compare rice profit per hectare between four groups (Table 6.5). Rice profit is an appropriate measure

TABLE 6.4
Comparison of Consumption Expenditure Per Capita
between Different Irrigation Statuses

(A) *Irrigation Status*	*Mean Value (St. Dev.)*	*Difference of the Means between Irrigation Status (A) and Irrigation status below (B)*		
		Mean of (A) – Mean of (B) (|t-ratio|)		
		(B)		
		(1) *Well Access Active Collective Management*	*(2)* *Well Access Inactive Collective Management*	*(3)* *No Well Access Active Collective Management*
1. Well access Active collective management	420 (223)			
2. Well access Inactive collective management	453 (209)	33 (0.78)		
3. No well access Active collective management	298 (179)	−122 (4.86)**	−155 (4.83)**	
4. No well access Inactive collective management	263 (165)	−157 (5.95)**	−190 (5.96)**	−35 (1.77)*

Notes: * significant at 10% level; ** significant at 1% level.

of the degree of overexploitation and resulting welfare loss. If groundwater becomes scarcer due to overexploitation, well users have to spend higher irrigation operation costs to acquire sufficient amount of irrigation water. Our filed observations indicate that the operation of valuable assets like pumps is carried out mostly by family labour. Therefore, we expect that well users' profit on rice after deducting family labour costs, in comparison with that of non-users, will not be appreciably large with overexploitation.

A key finding from this Table is that, in comparison with 'active collective management' villages (Groups 1 and 3), the profit of the farmers with 'well access' is not significantly different from that of the farmers with 'no well access' (difference of Rs 332 with *t*-ratio of 0.95).

TABLE 6.5

Comparison of Profit on Rice between Different Irrigation Statuses

		Difference of the Means between Irrigation Status (A) and Irrigation Status below (B)		
		Mean of (A) – Mean of (B) ($\|t\text{-ratio}\|$)		
		(B)		
		(1)	*(2)*	*(3)*
				No Well
		Well Access	*Well Access*	*Access*
	Mean	*Active*	*Inactive*	*Active*
(A)	*Value*	*Collective*	*Collective*	*Collective*
Irrigation status	*(St. Dev.)*	*Management*	*Management*	*Management*
1. Well access Active collective management	2274 (2964)			
2. Well access Inactive collective management	767 (2954)	–1507 (2.65)**		
3. No well access Active collective management	1942 (2522)	–332 (0.95)	1175 (2.60)**	
4. No well access Inactive collective management	–376 (2909)	–2650 (6.51)**	–1143 (2.16)**	–2318 (7.62)**

Notes: * significant at 10% level; ** significant at 1% level.

This indicates that the use of well irrigation does not guarantee higher profit anymore in active villages, although farmers had initially adopted it for higher profit. Even worse is the case of the farmers with 'well access' in 'inactive collective management' villages (that is, Group 2). The results indicate that once the collective management becomes inactive, profit on rice turns out to be only Rs 767 even among well users, which is significantly lower than not only, the profit of well users in active villages (the difference of Rs 1,507 with *t*-ratio of 2.65), but also the profit of non-users in active villages (the difference of Rs 1,175 with *t*-ratio of 2.60). This result possibly stems from the fact that the groundwater scarcity becomes more serious in inactive villages because the deteriorated tanks become unable to recharge the groundwater table through percolation. The findings above indicate that while the adoption of private wells increases yield, it does not result in an increased profit presumably due to overexploitation of groundwater.

Conclusion and Policy Implications

Facing recent rapid replacement of traditional irrigation systems by modern irrigation systems, this chapter investigates the efficiency of modern systems' water management and the impact of the decline in traditional systems. A key finding is that the dissemination of private wells results in double tragedies in that the non-well users suffer increased poverty due to the decline in collective management of tank irrigation systems; even the well users suffer reduced profit on rice due to the overexploitation of groundwater. If we look at the change of rice yield only, the dissemination of well irrigation systems seems to contribute a productivity improvement. However, our research finds that above mentioned negative effects do exist. We may call these negative effects tragedies because no individual has the incentive to correct them. Regarding the first tragedy, the negative effect on the non-well users is the one created by the well users' exit from the collective management. Since the well users do not suffer from the decline in collective tank management, they have no incentive to correct it. The second tragedy is a typical example of the 'tragedy of commons'. Since the root of the problem is the negative externality from one well user to other well users, no incentive mechanism exists among each of them. Without policy interventions, the correction of these tragedies is difficult.

The revitalization of collective management would solve the first tragedy as it enables non-well users to access to irrigation water from tanks. One possible support for the revitalization is the construction of lined channels by blocks of cement (Kajisa et al., 2007). This technology significantly reduces the labour required for maintenance work and also increases the availability of water by minimizing seepage. Even after experiencing the exit of well users from collective management, the maintenance work becomes feasible due to the remaining smaller number of tank users.

Charging an appropriate fee for electricity would contribute to preventing double tragedies. Under the present practice of free electricity for agricultural purposes, the number of electric pumps tends to be more than the social optimum, resulting not only in the overexploitation of groundwater but also in the decline in collective management. Not only the cost of power, but also the social cost due to the negative externality of groundwater overexploitation must be charged, albeit it may not be easy to estimate the social cost. In order to deter the dissemination of private wells, another possible policy may be to charge a sales tax on pump sets. Then, the government may use the revenues from electricity charge or from pump sales tax for tank revitalization projects. This transfer can be considered legitimate because the well users will receive indirect benefit from the revitalized tanks in that the water from the tank permeates to re-supply the groundwater table. The amount of indirect benefit could be estimated in collaboration with engineers, which we leave for our future research.

Notes

1. Examples of the former approach include Bardhan (2000) where he uses the index of quality of maintenance of distributaries and channels and Dayton-Johnson (2000) where he uses the conditions of canals as the proxy for the existence of collective action. Bardhan (2000) also uses the latter approach where he uses the number of conflicts and the frequency of rule violations among beneficiaries. Another example of the latter approach is Fujiie et al. (2005) where they measure cooperation in terms of the success or failure in organizing several water management related activities.
2. The need for channel cleaning varies across villages depending on the structure of the irrigation systems and their environment. Some villages require channel cleaning annually, while others may need it only biennially or even less frequently. The need for de-silting work arises much less frequently, usually once or twice in 10 years.
3. This inactive WUO dummy has a high correlation (correlation coefficient 0.75) with a dummy that becomes one when channel cleaning had not been conducted in the last three years, indicating the consistency of villagers' cleaning behaviour and the subjective evaluation of the overall status. The inactive WUO dummy is consistent also with water supply conditions. Among the inactive villages, according to the classification of our variable, 48 per cent of the villages claimed that the availability of tank water had worsened, whereas the corresponding percentage goes down to 29 per cent in the active villages. Moreover, in the active villages, even in those which claimed that the situation had worsened, the majority claimed that the reason was bad rainfall rather than the poor management of irrigation facilities, whereas this was reversed in the inactive villages.
4. The weights used for computation of adult equivalent household size are 0.5 for a child of age below 5 years, 0.73 for a child 6–10 years, 0.83 for a child 11–14 years, 0.83 for a female above 14 years, and 1.0 for males above 14 years (Rao, 1983). Household members living outside of the household because of work are excluded

but members living outside because of educational pursuit are included as members of the household, on the assumption that students receive financial support from the household.

5. The means of consumption expenditure are not statistically different, presumably due to the existence of consumption-smoothing mechanisms to some extent.

6. Use of the national poverty line of Rs 324 monthly per capita for 1993–94, instead of US$ 1, does not change the qualitative results. The same applies to the comparison of the consumption value.

7. This classification, however, entails two kinds of potential biases. First, under our definition of this dummy variable, the no-access group would include non-owner farmers who actually have access to well irrigation but choose not to use it because they have enough water from tanks or rainfall. These farmers presumably achieve yield and income as high as the farmers who use wells. The incorrect inclusion of them in the no-access group would result in the underestimation of the negative impact of non-access. Second, if high income farmers selectively became well owners, the impact of non-access would be overestimated. These potential biases must be resolved econometrically. Besides, we must provide for other possible determinants of yield and income/consumption. For more detailed statistical analysis by means of regression models, see Kajisa et al. (2006). The qualitative results of this paper and those of Kajisa et al. (2006) are essentially the same.

References

Bardhan, P. 2000. 'Irrigation and Cooperation: An Empirical Analysis of 48 Irrigation Communities in South India', *Economic Development and Cultural Change*, 48 (4): pp. 847–65.

Dayton-Johnson, J. 2000. 'Determinants of Collective Action on the Local Commons: A Model with Evidence from Mexico', *Journal of Development Economics*, 62: pp. 181–208.

Fujiie, M., Y. Hayami and M. Kikuchi. 2005. 'The Conditions of Collective Action for Local Commons Management: The Case of Irrigation in the Philippines', *Agricultural Economics*, 33 (2): pp. 179–89.

Fertilizer Association of India (various issues). *Fertilizer Statistics*. New Delhi.

Kajisa, K., K. Palanisami and T. Sakurai. 2006. 'Double Tragedies by Dissemination of Private Wells: Groundwater Overexploitation and Poverty in Tamil Nadu, India', *FASID Discussion Paper Series on International Development Strategies*, No. 2006-10-005, Foundation for Advanced Studies on International Development, Tokyo.

———. 2007. 'Effects on Poverty and Equity of the Decline in the Collective Tank Irrigation Management in Tamil Nadu, India', *Agricultural Economics*, Vol 36 (3), 2007, pp. 347–62.

Palanisami, K., 2000. *Tank Irrigation: Revival for Prosperity*. New Delhi: Asian Publishing Service.

Rao, V.K.R.V. 1983. *Food, Nutrition and Poverty in India*. New Delhi: Vikas Publishing House.

7

Agricultural Technology and Children's Occupational Choice in the Philippines

KEIJIRO OTSUKA, JONNA P. ESTUDILLO
AND YASUYUKI SAWADA

Introduction

Modern agricultural technology in rice production consisting of modern varieties (MVs) of rice, irrigation and knowledge-intensive improved farm practices—the so-called 'Green Revolution' technology—was introduced in the mid-1960s for increasing food production in the midst of high population growth on closed land frontiers in tropical areas of Asia. It is well known by now that the introduction of modern agricultural technology has resulted in a significant increase in rice production through the rise in yield and cropping intensity (Pingali et al., 1997). Many studies also show that the MV-irrigation tandem has a positive and significant impact on farm income (for example, Estudillo and Otsuka, 1999).[1] Moreover, the benefits of MV-irrigation are found to have accrued not only to the farmer cultivators but also to the landless agricultural workers, through the inter-regional migration of labour from unfavourable production areas where the demand for labour was low and stagnant, to favourable areas, where the demand for labour increased due to enhanced labour intensities associated with new the technology

(David and Otsuka, 1994). The benefits of MV-irrigation technology spread to urban consumers through lower paddy prices resulting from increased supply of rice in the market (Binswanger and Quizon, 1989).

While there have been a number of studies on the direct impact of modern agricultural technology—for example, on rice production and farm income—much less is known about its indirect impact on, among other things, household-schooling-investment decisions. Parental wealth and income is commonly believed to restrict a poor child from achieving a socially optimal level of schooling through binding credit constraints. An earlier study by Estudillo et al. (2004) found that the household's permanent income positively and significantly affects schooling progression of children. This indicates that the modern agricultural technology, coupled with the implementation of land reform and increased non-farm employment opportunities, has exerted a significantly positive impact on schooling investments by increasing the permanent income of households. Estudillo et al. (2006) also found that increased farm productivity increases the pawning value of land and, consequently, the incidence of pawning. Pawning revenues, in turn, were found to have been used to finance schooling investments in children.

In this chapter, we aim to explore the dynamic causations running from the adoption of modern agricultural technology to household income, schooling investment and occupational choice of children. We have three hypotheses. First, the adoption of modern agricultural technology increases farm income that enables the households to accumulate extra funds that can be used to finance schooling investments by relaxing the binding credit constraints. Second, the more educated children are those who are more likely to engage in non-farm labour employment, where returns to schooling are higher. Third, and finally, the income effects on schooling investment have increased over time because of the rising demand for schooling brought about by the development of the non-farm labour market.

In order to test these hypotheses, we used a rare panel data set collected from the same set of households in the rural Philippines in 1985, 1989, 2001 and 2004. We divided our sample children into three generations, namely, respondents and siblings, older children of respondents and younger children of respondents of school age. This is because parental investment decisions are not static but respond systematically to dynamic changes in socioeconomic environments affecting the returns to different factors of production. In particular, we would like to explore the changing importance of agricultural technology, land and human capital in determining household income and schooling investment decisions from the 1980s (1985 and 1989) to the 2000s (2001 and 2004).

This paper has four remaining sections. Section two presents the theoretical framework, whereas Section three describes the sample households and children. Section four identifies the factors affecting household decisions on child schooling and children's occupational choice. Finally, Section five concludes this chapter.

Theoretical Framework

Theoretical Approach

Productivity growth in agriculture is made possible primarily by modern agricultural technology. Then household income from farming rises, which generates extra funds for investments in child schooling, as is portrayed in Figure 7.1. Although we do not directly analyze it in this chapter, the efficiency gains arising from technological change in agriculture will lead to the development of non-farm sectors by providing cheap wage goods to urban workers, contributing to earnings or savings of scarce foreign exchange and releasing labour and other resources to non-farm sectors.[2] After children complete higher education, they prefer to find employment in the non-farm sector because the non-farm sector offers higher rates of returns to schooling. With greater involvement in the non-farm sector, household non-farm income increases and becomes an additional source of funding for child schooling.

Ideally, we would like to analyze sequentially (1) the effect of technology on farm incomes, (2) effects of both farm and non-farm income on schooling investment, (3) the effect of completed schooling on choice of job, and (4) the effect of schooling on non-farm income, within a simultaneous equation system. While income is expected to affect investments or changes in schooling, it is the completed years of schooling that affects the choice of job and the amount of labour income. Our approach is to analyze (3), that is, the effect of schooling on choice of job, while endogeneizing schooling levels by using, among other things, the time-invariant household characteristics. We use time-invariant characteristics related to permanent income or wealth,

FIGURE 7.1
**Hypothesized Interrelationships among Modern Agricultural Technology,
Schooling Investments and Non-farm Employment**

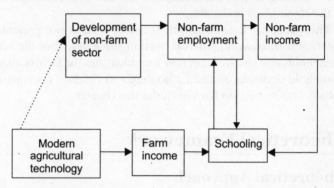

such as parental education, because schooling levels are determined by these long-term considerations. Next we analyze (1) and (4) by estimating farm and non-farm income functions at the household level and (3) by regressing incremental years in school of children of school age on the predicted farm and non-farm incomes.

Econometric Model

Probability to Engage in Non-farm Work

In order to explore the third issue discussed above, we specify a probit model of an individual choice to engage in non-farm work:

$$W = W(S, X), \qquad (1)$$

where W is a latent variable which represents the propensity of an individual to engage in non-farm work; S is years of schooling completed by the individual; and X is a set of other determinants of W. We expect S to exert a positive impact on W. For estimation purposes, we specify a linear specification for equation (1) as follows:

$$W = \alpha_0 + \alpha_1 E + \alpha_2 S + X\beta + u, \qquad (2)$$

where α_0, α_1, α_2, and β are the regression parameters and u is an error term. Since S is determined endogenously, we postulate a linear regression equation for S in which the variables capturing the child, parents and village characteristics are included among the explanatory variables. If the residuals of S and u follow a joint normal distribution with zero means, we can employ a two-stage simultaneous equation model suggested by Rivers and Vuong (1998).

Determinants of Schooling Progression of a Child

Here we specify a reduced form equation to describe the decision of parents to invest in a child in school:

$$\Delta E = E(Y^F, Y^N, X), \tag{3}$$

where E is the incremental years in school obtained by a child; Y^F and Y^N represent household farm income and non-farm income, respectively; and X is a set of the other determinants of ΔE. We expect that Y^F and Y^N have positive impacts on ΔE. We estimate equation (3) as a linear function as follows:

$$\Delta E = \gamma_0 + \gamma_1 Y^F + \gamma_2 Y^N + X\theta + e, \tag{4}$$

where γ_0, γ_1, γ_2 and θ are the regression parameters and e is an error term. Considering that Y^F and Y^N are determined endogenously, we postulate linear regression equations for these variables in which the variables capturing the impacts of modern agricultural technology, implementation of land reform and non-farm employment opportunities are included among the explanatory variables. Assuming that the residuals of Y^F and Y^N and e follow a joint normal distribution with zero means, we employ a two-stage simultaneous equation probit model proposed by Rivers and Vuong (1988).

Description of Sample Households

The Study Villages and Sample Households

We used a household-level panel data set collected from three rice-growing villages in the Philippines in 1985, 1989, 2001 and 2004.[3] These three villages may well be considered to represent irrigated, rain-fed, and upland ecosystems in the country. The irrigated and upland villages are located in Iloilo Province in Panay Island of Western Visayas and the rain-fed village is in Nueva Ecija Province in Central Luzon. These three villages were randomly selected from an extensive survey of 50 villages representing irrigated and lowland rice production environments in northern, central and southern Luzon as well as Panay Island (David and Otsuka, 1994).

The 1985 survey was a sub-sample of households selected from a complete enumeration of all households residing in the three villages based on a random sampling. The 1985 survey collected household-level information such as the adoption of modern agricultural technology, farm size, sources of household income and household asset holdings. The 1989 survey was a revisit of the 1985 sample and collected detailed information on member-specific characteristics such as age, schooling, land inheritance, occupational choice and children's enrolment in school on three generations of household members consisting of parents of the respondents, respondents and siblings and children of the respondents (Quisumbing, 1994). The 2001 and 2004 surveys included all households residing in the three villages and collected the same information as the 1985 survey. These latest two surveys also collected information on land market transactions for the purpose of exploring the linkages between land market transactions and parental investments in schooling (Estudillo, Sawada and Otsuka, 2006).

The sample households included both the farmer and landless agricultural worker households. Landless households engaged in rice farming are commonly observed to be the poorest in the village economy. They are employed in rice farming during peak labour demand such as transplanting and harvesting and, in more recent years, have been increasingly engaged in non-farm employment such as construction work, informal trade, housekeeping and laundry work. Farmer households included the share tenants, leaseholders and recipients of Certificate of Land Transfer (CLT) and Emancipation Patent (EP) as well as owner cultivators. EP and CLT holders are beneficiaries of the land reform. We included land under the EP in the category of owned land, as the EP confers full ownership right and also land under CLT in the category of leasehold land, as CLT holders pay amortization fees which are close to leasehold rent (Otsuka, 1991).

We have a total of 183 households in the benchmark survey in 1985 and 864 households in the latest survey in 2004 (Table 7.1). There was an increase in the proportion of landless households and a decrease in the average farm size because of population pressure on closed land frontier. Adoption of MVs was already as high as 83 per cent in 1985 and rose to 100 per cent in 2004. The proportion of area with irrigation

TABLE 7.1
Characteristics of Sample Households in the Study Villages
in the Philippines, 1985/89 and 2001/04

Characteristics	1985/89[1]	2001/04
Number of households	183	864
Farmer households (per cent)	75	46
Landless households (per cent)	25	54
Average farm size (ha)[1]	1.31	0.93
Adoption of modern technology:		
Modern varieties of rice (per cent area)	83	100
Irrigation (per cent area)	40	61
Sources of household income:		
Farm (per cent)	66	22
Non-farm (per cent)	34	78
Total annual household income (PHP nominal)	14,602	107,615
Total annual household income (PHP real)[2]	304	569

Notes: [1]Refers to farmer households only.
[2]Deflated by the provincial CPI using 1993 as the base year.

also increased because of the increase in the adoption of water pumps in the rain-fed village.

The proportion of household income derived from non-farm sources rose from 34 per cent in 1985 to 78 per cent in 2004. Such a sharp increase in non-farm income took place primarily because of the increased contribution to household income of children who are engaged in non-farm work, and also the rise in remittances received by the households from their members living outside the village including those abroad. Along with the shift in household income structure in favour of non-farm sources is the increase in real household income by nearly twice. It is clear that the increased contribution of non-farm income is the major force behind the increase in household income in 2004. Thus, it is critically important to analyze factors affecting the choice of non-farm jobs and the amount of non-farm income.

The Sample Children

We have a total of 1,046 individuals consisting of the respondents and siblings; 1,518 older children of the respondents, who are of age 22 years and above at the time of the survey; and 1,092 younger children of the respondents, who are in school age between seven and 21 years (Table 7.2). The parents of the respondents who were born around 1910 on average, had only slightly more than three years of

TABLE 7.2
Characteristics of the Children and Parents in the Study
Villages in the Philippines, 1985/89 and 2001/04

	1985/89	2001/04
Number of sample		
Respondents and siblings	1,046	NA
Older children of respondents[1]	NA	1,518
Younger children of respondents[2]	265	1,092
Average years of schooling completed:		
Parents of respondents:		
Father	3.6	NA
Mother	3.3	NA
Respondents and siblings:	6.7	NA
Older children of respondents	NA	10.4
Percentage engaged in non-farm work:		
Respondents and siblings	59	NA
Older children of respondents	NA	80
Percentage of children currently enrolled in school	69	70
Average inherited land (ha):		
Parents of respondents:[3]		
Father	2.74	NA
Mother	0.52	NA
Respondents and siblings	0.27	
Older children of respondents	NA	0.18

Notes: NA means not available.
  [1]Refers to children aged 22 years and above.
  [2]Refers to children aged between 7 and 21 years.
  [3]Refers to owned land.

schooling; the respondents and siblings had more than six; and the older
children of the respondents completed more than 10 years. The increase
in schooling attainment reflects increasing preference of parents
to invest in children's schooling, perhaps in response to the rising
returns to schooling because of the development of non-farm sector.

There has been an in creasing trend in children's involvement in
non-farm work: 59 per cent of the respondents and siblings were
engaged in non-farm work in 1989, whereas 80 per cent of older
children of respondents were able to do so in 2004. There has also
been a rise in the number of overseas workers: the number of overseas
workers increased from six people in 1985 to 71 people in 2004 in the
irrigated village, from five to 34 people in the rain-fed village,
and from six to 26 people in the upland village. According to
Estudillo et al. (2004), the years of schooling completed is by far

the most important factor affecting the decision to work abroad. A stronger preference of children for non-farm jobs and overseas work indicates that non-farm work has become more lucrative than farm work.

School enrolment rates of children remained at around 70 per cent in both 1989 and 2004. The primary school enrolment rate was close to 100 per cent because the three villages have free public elementary schools. Secondary school enrolment rate rose from 76 per cent in 1989 to 79 per cent in 2004, while tertiary school enrolment rate unexpectedly went down from 46 per cent to 35 per cent. While we believe that the importance of tertiary schooling in the country as a whole has gone up, the decline in enrolment rate may be simply because of the sample household composition—in 2004, we had more landless households, which were less able to send their children to tertiary schools.

Fathers of the respondents owned 2.74 ha of land while mothers only owned 0.52 ha. In the generation of the parents of respondents, women were worse off than men in terms of land as well as schooling attainment. In the respondent's generation, sons inherited about 0.50 ha while daughters inherited about half of that amount, which reflects the preference of Filipino parents to bequeath land to sons and send daughters to higher levels of schooling (Quisumbing et al., 2004). Overall, across generation of households, there has been a declining trend in the amount of inherited land because of the increasing scarcity of land.

The Determinants of Children's Schooling and Occupational Choice of Children

Probability to Engage in Non-farm Work

We specify a two-stage probit function to describe the choice of respondents and siblings and older children of respondents to engage in non-farm work by using the following independent variables: (I) characteristics of the child (age, age-squared, endogenous completed years in school); (II) inherited land of the child which can be under ownership, leasehold tenancy or share tenancy; (III) adoption of MV and irrigation in the previous survey year; (IV) the number of working age members, who are of age 22 years and above, in the previous survey; and (V) village dummies. The explanatory variables for completed years in school, which is treated as endogenous, are the following: (I) characteristics of the child (age, age-squared and daughter dummy); (II) characteristics of the parents (father's and mother's education and inherited land); and (III) village dummies.

As shown in Table 7.3, age affects completed years in school in a parabolic fashion reflecting a strong cohort effects. Daughters, in both 1989 and 2004, have completed more years in school than sons and such schooling gap becomes greater in 2004. Mother's education significantly increases children's completed years in school in both 1989 and 2004 and such effect is far greater than that of the father's schooling. Parent's inherited land which is assumed to reflect parental resource constraint, positively and significantly affects children's completed years in school: mother's inherited land in 1989 and father's inherited land in 2004 have positive and significant coefficients. Children in the irrigated village, which is closer to the town centre, have completed significantly more years in school than children in the upland village (control). Since access to land, availability of irrigation facilities and schooling are critical factors affecting household permanent income (or wealth), the above findings strongly indicate that permanent income is a major determinant of children's schooling.

Table 7.4 shows the results of the probability of individual children to engage in non-farm work. Children with higher completed years in school are much more likely to engage in non-farm work in both 1989 and 2004, which strongly indicates that returns to schooling are higher in non-farm work. Daughters are more likely to join the non-farm labour market than sons because Filipino daughters have a comparative advantage in non-farm work, while sons have a comparative advantage in farming (Quisumbing et al., 2004). It is interesting to find that the size of inherited land of the child negatively affects the probability to engage in non-farm work in 1989, but not in 2004. This result seems to indicate that when there were a relatively few attractive non-farm jobs, access to land affected the choice of farming decisively, but in more recent years, regardless of access to land, many people prefer non-farm jobs.

The characteristics of the household to which the children belong, such as the adoption of MVs and irrigation and the number of working age members, who are of age 22 years and above, do not seem to have exerted a significant impact on the choice of children to engage in

TABLE 7.3
The Determinants of Completed Years in School of Children in the Study Villages in the Philippines, 1989 and 2004 (Ordinary Least Squares)

Independent Variables	Respondents and Siblings 1989		Older Children of Respondents[1] 2004	
	Coef	t-values	Coef	t-values
Age in 1989 or 2004	0.06*	1.82	0.17**	3.76
Age-squared	−0.00	−0.36	−0.002**	−4.68
Daughter dummy	0.44**	2.28	0.82**	5.94
Father's education	0.03	0.78	0.19**	6.86
Mother's education	0.20**	3.98	0.29**	10.10
Father's inherited lands (ha)	0.01	0.38	0.34**	4.79
Mothers inherited lands (ha)	0.15**	2.39	−0.12	−0.72
Dummy for rain-fed village	−0.19	−0.74	−0.72**	−4.13
Dummy for irrigated village	0.88**	3.59	0.67**	3.73
Constant	8.91**	9.11	3.75**	4.09
Number of observation	1046		1518	
R-squared	0.21		0.27	

Notes: [1]Refers to children of age 22 years and above.
** = significant at 1 per cent level; * = significant at 5 per cent level.

TABLE 7.4
Determinants of the Probability of Children to Engage in Non-farm Work in the Study Villages in the Philippines, 1989 and 2004 (Probit)

Independent Variables	Respondents and Siblings 1989		Older Children of Respondents[1] 2004	
	Coef	z-values	Coef	z-values
Age in 1989 or 2004	−0.03*	−1.81	−0.07*	−1.75
Age-squared	0.0003**	2.52	0.0007	1.43
Education	0.18**	2.64	0.20**	2.70
Residual of the education function[2]	−0.07	−1.07	−0.09	−1.24
Daughter dummy	1.57**	15.16	0.90**	4.37
Inherited owned lands of the child (ha)	−0.56**	−4.17	0.09	0.43
Inherited leasehold lands of the child (ha)	−0.47**	−2.98	−0.53	−1.36
Inherited share tenancy lands of the child (ha)	−0.31**	−2.74	−1.07	−1.46

(*Table 7.4 Continued*)

(*Table 7.4 Continued*)

Independent Variables	Respondents and Siblings 1989		Older Children of Respondents[1] 2004	
	Coef	z-values	Coef	z-values
MV-irrigation ratio in 2001			0.05	0.29
Working members who are 22 years and above in 2001			−0.01	−0.32
Dummy for rain-fed village	−0.22	−1.96	0.26	1.18
Dummy for irrigated village	−0.04	−0.35	0.40*	1.95
Constant	−0.78	−0.89	−0.03	−0.03
Number of observation	1045		401	
Log-likelihood	−488.82		−155.73	

Notes: [1]Refers to single children of age 22 years and above.

[2]Residuals obtained from the Ordinary Least Squares estimates of completed years in school in Table 2.

* = significant at 5 per cent level; ** = significant at 1 per cent level.

non-farm work. These findings indicate that individual characteristics have exerted a far greater influence on the choice of the individual to engage in non-farm work. Children in an irrigated village, which has an easy access to the towns, are more likely to engage in non-farm work.

Overall, we have found strong evidence supporting our second hypothesis that the more education children are more likely to engage in non-farm labour employment. Non-farm work has become more attractive to children in more recent years, regardless of their access to land, most likely because non-farm work offers a higher rate of returns to schooling.

In the Philippines as a whole, the percentage of rural non-farm employment was 35 per cent in 1983; it increased to 68 per cent in 2003 (Estudillo et al., 2006). The rural labour force has been moving out of farming because non-farm activities have become a more profitable economic endeavour. It is commonly observed that the younger and the more educated members of the households are those who are more likely to be involved in non-farm activities. Females compared to males have the higher propensity to engage in non-farm work: the proportion of female labour force engaged in non-farm work was more than 80 per cent in 2000 and 2003.

The most common forms of non-farm employment are the rural industries such as ready-made garments as well as metal craft, plastic, paper and wood craft. There was a spark of a labour-intensive rural industrialization in the Philippines in the late 1970s and the early 1980s, which, according to Hayami et al. (1998, p. 152), was strengthened by the progress in liberalization and de-regulation in trade, foreign exchange, and foreign direct investment policies since the mid-1980s. Some other major non-farm jobs, especially, those in services sectors, are commonly found in informal sector such as transport, domestic and personal services, trade, and communication.

It is also important to mention the rapid growth in the number of overseas Filipino workers (OFWs): the number of OFWs in the Philippines rose from 372,000 in 1985 to 1,056,000 in 2004. The common destinations are Saudi Arabia for males and Hong Kong, Singapore, Japan and China-Taipei for females. The evolution of labour markets in rural industries and services and the rise in the number of OFWs indicate that the rural labour market has been becoming integrated with the larger domestic and international labour markets.

Schooling Progression of Children in School Age

We estimate a two-stage probit model of the determinants of schooling progression of children defined as the incremental years in school between 1985 and 1989 and between 2002 and 2004. We use the following sets of independent variables: (I) Endogenous household farm and non-farm incomes and their corresponding residuals from the first-stage income functions; (II) characteristics of the parents such as father's and mother's education and father's and mother's inherited land; (III) characteristics of the child such as age and age-squared and daughter dummy; and (IV) village dummies.

The explanatory variables for farm and non-farm incomes in the first-stage regressions are: (I) adoption of MVs and irrigation; (II) land size by tenure; (III) the number of household members by age group, that is, members between 22 to 30 years, between 31 to 40 years, between 41 to 50 years, between 51 and 60 years, and 61 years and above; (IV) ratio of female members; (V) ratio of members with secondary and tertiary schooling; (VI) ratio of sick members; (VII) dummy for ownership of carabao, tractor and thresher; and (VII) village dummies.[4]

Table 7.5 shows the determinants of farm and non-farm incomes separately in 1985 and 2004. The adoption of modern rice varieties

TABLE 7.5
The Determinants of Farm and Non-farm Income in the Study Villages in the Philippines, 1985 and 2004 (Ordinary Least Squares)

Variables	1985				2001			
	Farm Income		Non-farm Income		Farm Income		Non-farm Income	
	Coef	t-values	Coef	t-values	Coef	t-values	Coef	t-values
MV* Irrigation[1]	−1669	−1.01	626	0.27	−6224	−1.62	−22599**	−2.20
Owned and EP lands (ha)[1]	10944**	7.70	−1023	−0.51	25075**	8.44	4395	0.55
Leasehold and CLT lands (ha)[2]	5693**	7.22	36	0.03	37500**	15.41	2048	0.31
Share tenancy lands (ha)	3873**	2.60	−2123	−1.02	17491**	3.47	−315	−0.02
Number of members:								
Between 22 to 30 years old	1694*	1.93	−821	−0.67	885	0.46	−10311*	−1.99
Between 31 to 40 years old	1931*	1.76	−1454	−0.94	1127	0.50	3069	0.51
Between 41 to 50 years old	2648**	2.07	257	0.14	5494*	1.79	22457**	2.73
Between 51 to 60 years old	2953**	2.60	−1015	−0.64	−7499**	−2.86	26720**	3.79
61 years old and above	1664	1.44	3678**	2.26	−2489	−1.04	7246	1.13
Ratio of female members	1318	0.33	3231	0.58	−4864*	−0.52	−4767	−0.97
Ratio of working members with:								
Secondary schooling	5097**	2.20	3844	1.18	−1382	−0.22	49062**	2.88
Tertiary schooling	2850	0.83	4353	0.91	17715**	2.35	144917**	7.18
Ratio of sick members	6534	1.08	9089	1.07	−27302	−1.34	−2383	−0.04

Dummy for carabao ownership	−1385	−0.89	−1536	−0.70	−3608	−0.71	−16557	−1.22
Dummy for thresher ownership	−253	−0.05	−2259	−0.35	1761	0.12	−49046	−1.28
Dummy for tractor ownership	7459**	2.61	1680	0.42	24996**	4.42	16898	1.11
Dummy for rain-fed village	1451	0.75	−1227	−0.45	2604	0.52	18675	1.40
Dummy for irrigated village	5346**	2.62	3870	1.35	5987	1.16	46320**	3.34
Constant	−6569**	−2.13	3047	0.70	9312	1.39	−5162	−0.29
Number of observations	182		182		726		726	
R-squared	0.55		0.14		0.39		0.15	

Notes: ¹Refers to Emancipation Patent.

²Refers to Certificate of Land Transfer.

* = significant at 5 per cent level; ** = significant at 1 per cent level.

and irrigation did not seem to have had a significant impact on farm income in both 1985 and 2001 because the adoption of MVs reached high levels even as early as 1985 and reached 100 per cent by 2001. In addition, the adoption of modern rice varieties increased farm income even in the rain-fed condition, but more importantly, due to the fact that irrigation effect is captured by irrigated village dummy. In all likelihood, the irrigated village has a significantly higher farm income than the upland village (control) in 1985 because of the 100 per cent adoption of MVs and complete irrigation. In brief, evidence seems to support the hypothesis that modern agricultural technology significantly increases farm income, which may become an important source of funds to finance children's schooling.

Farmer households such as owner cultivators, leaseholders, and share tenants have higher farm income than landless households (control), implying that access to land is a major determinant of income in rural households. It is also interesting to find that the number of working members aged between 22 to 30 years, between 31 to 40 years, between 41 to 50 years and between 51 to 60 years, significantly increased farm income in 1985, indicating that a larger number of household labour force was absorbed in agriculture in 1985. The situation was different in 2001 when only members aged between 31 to 40 years significantly increase farm income. A majority of the household labour force was involved in non-farm activities in 2001 because such activities became more lucrative vis-a-vis farming. This might been brought about partly by the decline in international and domestic rice prices due to the spread of Green Revolution to South Asia and partly by the development of non-farm labour markets. It may be noted that since we have controlled for the effects of schooling, an increase in non-farm income due to the increase in the number of working household members implies that the development of non-farm sectors has provided ample employment opportunities to most age groups regardless of educational qualifications. To the extent that the non-educated workers are children of poor families, the present evidence indicates that the development of non-farm sectors is pro-poor.

In 2001, it was seen that female members had a lower propensity to choose farm jobs as is shown by the negative coefficient of the ratio of female members in the farm income function. Filipino females tend to be more educated than males owing to parental preferences to send females to higher levels of schooling. Secondary and tertiary schooling increased non-farm income significantly in 2001 and the effect of tertiary schooling was far greater than that of secondary schooling. In 1985, because of the limited non-farm labour employment opportunities, members with secondary schooling were involved in farming, as shown by a positive and significant coefficient of secondary schooling in the farm income regression. But this was no longer the case in 2004, when the non-farm labour market had expanded, so that household members with secondary and tertiary schooling tended to be engaged in non-farm work. It seems that the less educated such as those with primary schooling only are left to farming, as schooling level does not affect farm efficiency in the Philippines (David and Otsuka, 1994; Quisumbing et al., 2004).

Table 7.6 shows the results of the probit model of the determinants of incremental years in school of children in school age. Farm income significantly increased children's progress in school in 1985, but not so in 2004, when non-farm income exerted a positive and significant impact. These results show the increased importance of non-farm income in financing schooling investments of young children in as much as non-farm income comprises the largest portion of total household income in 2004. It is also important to point out that both father's and mother's education, significantly, increased the children's progress in school in 2002–04. Since parents' education is a critically important variable in the income functions (Table 7.5), the significant effect of the schooling variables would capture, at least partly, the income effect on children's schooling enrolment. Thus, the hypothesis that income is a major determinant of schooling investment cannot be rejected. Mother's inherited land significantly increased children's progress in school in 1985 indicating the important role of the land pawning market in financing schooling investments. Inherited lands of parents, however, decreased their importance in 2004 perhaps because of the increase in household non-farm income. Age and age-squared are significant, indicating that school progression increases in a parabolic fashion with children's age. Children in rain-fed village are significantly less likely to progress in school compared to children in the upland village (control).

Overall, our regression results give evidence that modern agricultural technology significantly increases farm income, which, in turn, significantly affects children's progress in school. The fact that school progress is significantly affected by household income indicates that

TABLE 7.6
The Determinants of the Incremental Years in School of
Children in the Study Villages in the Philippines, 1985 and 2004

| Variables | Children of Respondents | | | |
| | 1985–89 | | 2002–2004 | |
	Coef	z-values	Coef	z-values
Farm income				
(in thousand PHP)	8.39**	3.07	0.24	1.59
Residual of farm income function	−4.06	−1.22	−0.06	−0.35
Non-farm income				
(in thousand PHP)	−10.03	−1.27	0.34**	2.31
Residual of non-farm income	6.85	0.94	−0.24	−1.58
Father's education	−0.04	0.53	0.03**	2.25
Mother's education	0.11	1.30	0.03**	2.50
Fathers inherited lands (ha)	0.01	1.52	0.00	0.05
Mother's inherited lands (ha)	0.95*	1.73	−0.08	−0.57
Daughter dummy	0.43	1.14	0.18**	2.23
Age of the child	1.15**	5.27	0.49**	4.72
Age-squared	−0.06**	−6.95	−0.02**	−7.06
Dummy for rain-fed village	−0.92*	−1.95	−0.97**	−9.05
Dummy for irrigated village	0.65	0.81	−0.23	−1.62
Constant	−1.92	−1.47	−0.34	−0.42
Number of observations	248		870	
Log-likelihood	−367.40		−1210.94	

Notes: * = significant at 5 per cent level; ** = significant at 1 per cent level.

schooling remains a superior good in the rural Philippines as was found earlier in Vietnam (Behrman and Knowles, 1999).

Structural Interpretation

How can we interpret the nexus between increased farm income and enhanced child schooling structurally? Estudillo et al. (2006) demonstrate that the emergence of an informal land market has played a key role in facilitating investments in children's schooling. Indeed, *sangla* has become common over time. *Sangla* is a credit contract where the farmer temporarily transfers his cultivation rights on a farmland to the pawnee in exchange for cash with an agreement to redeem it upon loan repayment without interest charges. Considering that the land productivity has improved significantly due to the adoption of modern agricultural technology over time, pawning fees under *sangla* contract have increased. Accordingly, accessibility to credit markets should have improved as land pawning becomes popular. In all likelihood, pawning has become an important source of schooling investment over time by decreasing the households' credit constraints.

The informal land market, however, is not accessible to the landless, but nonetheless these households have benefited as well; this is because the recipients of pawned out lands, who are not familiar with rice farming, commonly hire in landless labour either on a casual daily arrangement or as seasonal permanent worker (Hayami and Otsuka, 1993). In brief, it is possible that the development of an informal land market has strong linkage with investments in schooling.

In order to investigate the role of land pawning in facilitating credit accessibility which leads to schooling investments we estimate a binary dependent variable model of credit constraints by following Jappelli (1990). We define an indicator variable of credit constraint which takes the value of either unity, if the credit constraint is binding, or zero, if otherwise. While typical household surveys do not collect credit information that enables identification of the credit condition directly (Scott, 2000), we designed the credit module in our questionnaire carefully so that we can identify credit-constrained households directly. We estimate the probit model under the normality assumption.

Table 7.7 presents estimated results of the credit constraints model with the following as explanatory variables: (1) ratio of pawned land to total land; (II) ratio of household members of age 22 years and above, ratio of members with secondary schooling, ratio of members with tertiary schooling, ratio of members who are sick, and ratio of overseas workers; (III) total household income and its squared term; (IV) farm size and its squared-term; and (V) number of *carabao* (water buffalo).

Once the ratio of pawned land is treated as an endogenous variable by including its residual from the first-stage regression, as suggested by Rivers and Vuong (1988), its coefficient becomes negative and statistically significant, indicating that more land is pawned out and the household faces less credit constraints. This finding is consistent with the process that the Green Revolution and the implementation of land reforms induce the evolution of an active land pawning market for farmlands which enables parents with access to farmlands to use the pawning to finance high pay-off long-term investments in schooling.

TABLE 7.7
Determinants of Credit Constraints in the Study Villages in the Philippines, 2001

Estimation Method		Probit	Probit	Probit	IV probit	IV probit	IV probit
Residual from the first stage regression	respawn 3				1.177* (1.890)	1.465** (2.220)	1.275* (1.900)
Ratio of pawned land to total land (endogenous variable)	Rpawn	0.093 (0.580)	0.082 (0.520)	0.107 (0.650)	−1.002* (1.670)	−1.289** (2.020)	−1.090* (1.670)
Number of members (22 years old and above)	adult 22	0.024 (0.410)	0.027 (0.440)	0.029 (0.480)	0.032 (0.540)	0.039 (0.650)	0.039 (0.650)
Ratio of members with secondary schooling	Rsec		0.173 (0.880)	0.158 (0.800)		0.128 (0.650)	0.122 (0.620)
Ratio of members with tertiary schooling	Rcoll		−0.047 (0.190)	−0.116 (0.470)		−0.203 (0.800)	−0.240 (0.940)
Ratio of female members	Rfemale		−0.190 (0.470)	−0.261 (0.640)		−0.308 (0.770)	−0.350 (0.870)
Ratio of sick members	Rsick		−1.341* (1.700)	−1.245 (1.610)		−1.307* (1.680)	−1.234 (1.610)
Ratio of overseas workers	Rofw		−0.423 (0.860)	−0.373 (0.760)		−0.604 (1.200)	−0.541 (1.080)
Total income	totincom	−1.17E−06 (1.27)	−9.91E−07 (0.98)	−1.46E−06 (1.39)	−1.50E−06 (1.59)	−1.17E−06 (1.14)	−1.56E−06 (1.44)
Total income squared	totincom 2	1.10E−12* (1.87)	1.04E−12 (1.6)	1.35E−12* (1.76)	1.34E−12** (2.1)	1.23E−12* (1.69)	1.49E−12 (1.61)

	(1)	(2)	(3)	(4)	(5)	(6)
Farm size	−0.163 (1.100)	−0.180 (1.190)	−0.181 (1.210)	−0.144 (0.960)	−0.153 (1.000)	−0.156 (1.040)
Farm size squared	0.023 (0.790)	0.028 (0.950)	0.031 (1.060)	0.021 (0.730)	0.026 (0.880)	0.028 (0.980)
Number of carabao			−0.284* (1.970)			−0.233 (1.580)
Constant	0.162 (0.860)	0.224 (0.790)	0.393 (1.330)	0.260 (1.340)	0.429 (1.460)	0.540* (1.790)
Number of observations	381	381	381	381	381	381

Notes: * = significant at 10 per cent level; ** = significant at 5 per cent level; *** = significant at 1 per cent level.

Conclusions

This study explores how modern agricultural technology has affected parental decisions to invest in child schooling and the subsequent decision of children to join the non-farm labour market. Our major finding is that modern agricultural technology increases farm income of rural households, which generates funds to finance children's schooling by relaxing binding credit constraints. Another important finding is that the more educated children are found to have a higher probability to join the non-farm labour market, where they can obtain higher rates of returns to their schooling. Participation in the non-farm labour market increases household non-farm income and remittances, which, in turn, contribute to poverty reduction and stimulate further schooling investments.

Given the experience in the Philippines, it is reasonable to argue that modern agricultural technology is a major catalyst of rural development, as it strengthens household food security, increases household income, relaxes the binding constraints on credit and, hence, induces investments in schooling of children (Barker and Dawe, 2002). T.W. Schultz (1980) once argued that the decisive factors to improve the welfare of poor rural people are not space, energy, and cropland but they are the improvements in human quality in terms of health and schooling and scientific knowledge. Our study reinforces his arguments by providing the evidence that agricultural technology decreases the incidence of poverty not only by increasing income in the short run, but also by nurturing and sending an educated labour force to non-farm sectors in the long run. Such inter-sectoral labour transfer will certainly stimulate the development of non-farm sectors and contribute to the alleviation of rural poverty by generating additional income and remittances.

Acknowledgements

The authors thank the International Rice Research Institute (IRRI) for permission to use the 1985 data set and Agnes R. Quisumbing and Geetha Nagarajan for the 1989 data set. The IRRI and the Foundation for Advanced Studies on International Development in Tokyo, Japan have provided generous funding for the survey in 2001 and 2004. Jessaine Soraya C. Sugui processed the 2004 data. The usual caveat applies.

Notes

1. See an introduction to a special issue on 'the role of non-farm income in poverty reduction' by Otsuka and Yamano (2006) and other papers included in the November 2006 issue of *Agricultural Economics*.
2. Also, the development of non-farm sector may be induced by the adoption of modern agricultural technology through the consumption and production linkages (for example, Delgado et al., 1998).
3. There were also surveys in 1993 and 1997, but we did not include these surveys in this chapter. See Estudillo, Sawada and Otsuka (2006) for more details on the description of the study villages and sample households.
4. This specification is drawn from Estudillo et al. (2006).

References

Behrman, Jere R. and James C. Knowles. 1999. 'Household Income and Child Schooling in Vietnam'. *World Bank Economic Review,* 13(2): pp. 211–56.

Binswanger, Hans P. and J.B. Quizon. 1989. 'What Can Agriculture Do for the Poorest Rural Group?', in I. Adelman and S. Lane (eds), *The Balance between Industry and Agriculture in Economic Development.* Hampshire, UK: Macmillan Press.

David, Cristina C. and Keijiro Otsuka. 1994. *Modern Rice Technology and Income Distribution in Asia,* Boulder, CO: Lynn Rienner.

Delgado, Christopher, Jane Hopkins, Valerie A. Kelly, Peter Hazell, Anna A. McKenna, Peter Gruhn, Behjat Hojjati, Jayashree Sil and Claude Courbois. 1998. 'Agricultural Growth Linkages in Sub-Saharan Africa', International Food Policy Research Institute Research Report 107, Washington, D.C.

Estudillo, Jonna P. and Keijiro Otsuka. 1999. 'Green Revolution, Human Capital, Off-farm Employment: Changing Sources of Income among Farm Households in Central Luzon, 1966–94', *Economic Development and Cultural Change,* 47(3): pp. 497–523.

Estudillo, Jonna P., Tetsushi Sonobe and Keijiro Otsuka. 2006. 'The Development of the Rural Non-farm Sector in the Philippines and Lessons from the East Asian Experience', in A.M. Balisacan and H. Hill (eds), *The Dynamics of Regional Development: The Philippines in East Asia.* Cheltenham, UK: Edward Elgar.

Estudillo, Jonna P., Yasuyuki Sawada and Keijiro Otsuka. 2004. 'The Determinants of Schooling Investments of the Rural Filipino Households, 1985–2002', *Philippine Review of Economics,* 41(1): pp. 1–29.

Estudillo, Jonna P., Yasuyuki Sawada and Keijiro Otsuka. 2006. 'The Green Revolution, Development of Labor Markets, and Poverty Reduction in the Rural Philippines, 1985–2004', *Agricultural Economics*, 35, Supplement: pp. 399–407.

Hayami, Y., M. Kikuchi and E. Marciano. 1998. 'Structure of Rural-based Industrialization: Metal Craft Manufacturing on the Outskirts of Greater Manila, Philippines', *The Developing Economies*, 26(2): pp. 132–54.

Hayami, Yujiro and Keijiro Otsuka. 1993. *The Economics of Contract Choice: An Agrarian Perspective*. New York: Oxford University Press.

Jappelli, Tullio. 1990. 'Who is Credit Constrained in the U.S. Economy?', *Quarterly Journal of Economics*, 105(1): pp. 219–34.

Otsuka, Keijiro. 1991. 'Determinants and Consequences of Land Reform Implementation in the Philippines', *Journal of Development Economics*, 35(2): pp. 339–55.

Otsuka, Keijiro and Takashi Yamano. 2006. 'Introduction to the Special Issue on the Role of Non-farm Income in Poverty Reduction: Evidence from Asia and East Africa', *Agricultural Economics*, 35(3): pp. 393–97.

Pingali, P.L., Mahabub Hossain and R.V. Gerpacio. 1997. *Asian Rice Bowls: The Returning Crisis?* Wallingford, UK: CAB International.

Quisumbing, Agnes R. 1994. 'Intergenerational Transfers in Philippine Rice Villages: Gender Differences in Traditional Inheritance Customs', *Journal of Development Economics*, 43(2): pp. 167–95.

Quisumbing, Agnes R., Jonna P. Estudillo and Keijiro Otsuka. 2004. *Land and Schooling: Transferring Wealth across Generations*. Baltimore, MD: John Hopkins University Press.

Rivers, D. and Q. H. Vuong. 1998. 'Limited Information Estimators and Exogeneity Tests for Simultaneous Probit Models', *Journal of Econometrics*, 39, pp. 347–66.

Schultz, Theodore W. 1980. 'Nobel Lecture: The Economics of Being Poor', *Journal of Political Economy* 88(4): pp. 639–55.

Scott, Kinnon. 2000. 'Credit', in Margaret Grosh and Paul Glewwe (eds), *Designing Household Survey Questionnaires for Developing Countries: Lessons from Ten Years of LSMS Experience*. Washington, D.C.: The World Bank.

8

A Panel Data Model for the Assessment of Farmer Field Schools in Thailand

SUWANNA PRANEETVATAKUL AND HERMANN WAIBEL

Introduction

Projects on farmer training in Integrated Pest Management (IPM) in developing countries using the Farmer Field School (FFS) approach are widely implemented by donor organizations, including the World Bank. This is in spite of criticism that such projects are fiscally unsustainable (Quizon et al., 2001) and are not always effective in changing pest management practices or in improving farm performance (Feder et al., 2003) and have only limited diffusion effects (Rola et al., 2002; Feder et al., 2004). On the other hand, it was shown that FFS could improve farmer knowledge in pest identification and improve their ecosystems understanding (Godtland et al., 2004; van den Berg, 2004 and Tripp et al., 2005). Also, it was found that public investments in integrated pest management programmes in cotton in Asia showed good rates of return (Erickson, 2004; Ooi et al., 2005). Moreover, in China, where bollworm-resistant transgenic cotton varieties have been widely introduced, FFS was found to be effective in helping farmers realize the potential of pesticide reduction that Bt varieties offer (Yang et al., 2005).

A common facet of past impact analyses of Farmer Field School projects was that the data used did not allow the definition of good counterfactual scenarios because no control area was available, or only insufficient baseline data existed. Also, comparisons were based on only two observation points before and after the training. In addition, most of these studies concentrated on simple performance parameters like knowledge, pesticide use and yield but did not include, for example, the impact on the environment. In this chapter, we use a set of panel data collected over a period of four years covering a maximum of eight rice-growing seasons from three groups of farmers. The analysis presented here is an advancement of an earlier study that analyzed the short-term impact of FFS in Thailand (Praneetvatakul and Waibel 2003).

Data and Impact Indicators

Data was collected in five pilot sites of the Department of Agricultural Extension (DOAE) in Thailand. In each pilot site, a Farmer Field School following the usual methodology, with a season long experiential training in the field (Kenmore, 1996), was implemented. The sample included 241 farmers and was composed of three groups: (1) training participants (FFS farmers), on the average 20 farmers per FFS; (2) 15 randomly selected non participant farmers per village but exposed to the FFS knowledge because they were living in the FFS village, (non FFS); (3) 15 unexposed farmers, randomly selected from a control village located near an FFS village (control farmers). The control villages had similar socioeconomic and natural production conditions but there was a low probability of intensive information exchange with the respective FFS villages. For example, the control villages had different market places than the FFS villages. The farmers were interviewed at three different points of time: (1) in February 2000, at the end of the wet rice-cropping season, which was before the training had started; (2) in February 2001, in the rice growing season after the training, that is, where farmers could apply their new knowledge for the first time; and (3) in February 2003, two years after the second survey. Thus, trained farmers in the five pilot villages had the opportunity to apply their new knowledge between four to eight rice growing seasons after the training, depending on the intensity of rice production. Unfortunately for the third survey, the sample size had to be reduced because of heavy flooding in two FFS villages so that some farmers in the panel could not harvest rice in the survey season.

The questionnaire included information on farm household characteristics, farmer knowledge on rice pest management, data on rice-production inputs and outputs and questions on health issues related to pesticide use. Particular emphasis was given to a detailed account of pesticide use regarding quantity, common and brand names, active ingredients and formulation.

To assess the impact of FFS, we defined several impact indicators. First, we measured farmers' knowledge of rice and pest management. A score was constructed from a set of knowledge questions developed in cooperation with national IPM experts. Second, total rice yields per farm, including sales and home consumption, were based on farmers' estimates and divided by the respective area planted to rice. Third, the amount spent on pesticides including insecticides, molluscisides (chemicals used to kill snails), fungicides and herbicides were calculated in $ per ha. Fourth, the gross margin of rice production in $ per ha, measured as total revenue above total variable costs excluding the value of family labour. Fifth, as a measure of farmer net benefit, we deducted health costs from chemical pesticide use from the gross margin. Since no data on occupational health was collected in this study, health costs were accounted for by using a ratio of pesticide costs to health costs of 1:1 based on the results of a study by Rola and Pingali (1993). In addition, as a non-monetary measure, the Environmental Impact Quotient (EIQ) was calculated to quantify the environmental and human toxicity effects of pesticides (Kovach et al., 1992). The EIQ index differentiates pesticide use according to crop type, pesticide type and quantity used as well as their toxicity to pesticide applicators, toxicity to consumers and toxicity to the ecology. The index sums up all negative side effects of pesticides; hence a higher EIQ number indicates a higher risk to health and environment.

The EIQ index can be calculated by the following formula (Kovach et al., 1992):

$$EIQ = \{[C \times (DT \times 5) + (DT \times P) + C \times (S+P)/2) \times SY+L+ (F \times R) + [(D \times (S+P)/2 \times 3] + (Z \times P \times 3) + (B \times P \times 5)\}/3$$

where:

C = chronic toxicity, DT = acute toxicity to human, P = half-life on the plant, L = ability of elution and penetration to the underground water, S = time of half-life in the soil, SY = penetration ability to the soil, F = toxicity to the aquatic species, R = surface lost ability, D = toxicity to birds, Z = toxicity to bees, and B = toxicity to beneficial insects.

Hence EIQ can be described as EI farmer + EI consumer + EI ecology, whereas:

$$\text{EI farmer} = C \times (DT \times 5) + (DT \times P).$$
$$\text{EI consumer} = C \times (S + P)/2) \times SY + L$$
$$\text{EI ecology} = (F \times R) + (D \times (S + P)/2 \times 3) + (Z \times P \times 3) + (B \times P \times 5)$$

With the existing information, the EIQ of almost any pesticide in a defined cropping system can be calculated and available in Kovach et al. (1992), whereby higher EIQ indicate high risks to environment and human health.

The Model

The analysis applies a difference in difference (DD) model (Greene 2000). DD models can be used to analyze changes in farm performance, due to a treatment, which can be attributed to external interventions such as farmer trainings. In our analysis, we proceeded in two steps. First we investigated linear shifts in performance and second we measured change as a growth process. The linear shift implies a one-off performance change at the observation point relative to the baseline period. The change in the growth rate takes account of the fact that the development process influences performance and assumes an exponential path in the rate of change of performance for trained and untrained farmers. Hence, the model accounts for the fact that change is taking place even without the FFS training. The linear shift was measured by applying a paired t-test (Anderson, Sweeney and Williams 2002), to test the differences between before and after training for FFS, non-FFS and control farmers. For those performance indicators where we find a significant linear shift, we proceed with the two and three period growth model. The rationale for this procedure is that we do not expect significant results as we increase the degree of rigour in the testing procedure, that is, if we do not get a significant difference in the t-test, a significant coefficient in an econometric growth model is unlikely. Since we have three observation points, over time we can apply two alternative models: a two-period and a three-period panel data model. With the three-period model, a simultaneous estimation of the time period effects is achieved using a larger sample.

In applying this model, we expand the procedure described in Feder et al. (2003) and used to measuring impact of IPM in Indonesia. Accordingly, the change in farmers' performance (for example, the yield) through training can be modelled as an exponential growth process. This is displayed in equation 1:

$$Y_1 = Y_0 \cdot e^{\{\alpha + \beta D_{nffs} + \mu D_{ffs} + \gamma \Delta X + \delta \Delta Z\}} \tag{1}$$

where:

Y_1	:	rice yield after the training,
Y_0	:	rice yield before the training,
α	:	coefficient for yield growth before the training,
μ	:	rate of yield growth of FFS farmers after the training,
β	:	rate of yield growth rate for the non-FFS farmers after training,
D_{ffs}	:	dummy variable for FFS farmers, for FFS = 1 and zero = otherwise,
D_{nffs}	:	dummy variable for non-FFS farmers, for non FFS = 1 and zero = FFS and control,
X	:	vector of farmer characteristics,
Z	:	vector of village characteristics,
γ and δ :		corresponding coefficients of these vectors,
Δ	:	the differencing operator between, before and after the training,
e	:	the exponential operator.

The specification for an empirical estimation of the model can be obtained by taking the natural log of equation (1) and rearranging it accordingly:

$$\Delta(\ln Y) = \alpha + \beta D_{nffs} + \mu D_{ffs} + \gamma \Delta X + \delta \Delta Z \tag{2}$$

where: $\qquad \Delta(\ln Y) = (\ln Y_1 - \ln Y_0)$

Multi-period Panel Data Model

Unlike in models that are based on cross sectional data, panel data allow for the unobserved effects, a_i, to be correlated with the explanatory variables (Wooldridge, 2000). This is because a_i is assumed to be constant over time, hence one can compute the difference in the observed parameters over the two years.

The equations for period 2 (eq 3) and period 1 (eq 4) are as follows:

$$Y_{i2} = (\delta_0 + \alpha) + \gamma_2 X_{i2} + a_i + u_{i2} \tag{3}$$

$$Y_{i1} = \delta_0 + \gamma_1 X_{i1} + a_i + u_{i1} \tag{4}$$

Subtracting the equation (4) from equation (3) results:

$$\Delta Y_i = \alpha + \gamma \Delta X_i + \Delta u_i \tag{5}$$

where:

Δ denotes the change from period 1 ($t = 1$) to period 2 ($t = 2$), Y_i is the dependent variable, X_i are independent variables and u_i is the error term. The unobserved effect, a_i, does not appear since it has been differenced away. The resulting intercept (α) denotes the change in the intercept between the two periods. γ is the coefficient of the change in X_i.

Extending the analysis to three periods ($t = 1, 2,$ and 3), the procedure is analogous as shown in equation (6):

$$Y_{it} = \delta_1 + \delta_2 d2_t + \delta_3 d3_t + \gamma_1 X_{it1} + \dots + \gamma_k X_{itk} + a_i + u_{it} \tag{6}$$

Equation (6) includes dummies for two periods, $d2$ and $d3$. The intercept for the first period is δ_1; for the second period it is $\delta_1 + \delta_2$. For period three, the definition of intercept is analogous. In the $t = 3$ case, time period one is subtracted from time period two and time period two from time period three resulting in Equation 7:

$$\Delta Y_{it} = \delta_2 \Delta d2_t + \delta_3 \Delta d3_t + \gamma_1 \Delta X_{it1} + \dots + \gamma_k \Delta X_{itk} + \Delta u_{it} \tag{7}$$

for $t = 2$ and $t = 3$. Equation (7) contains the differences in the time period dummies, $d2_t$ and $d3_t$; that is, for $t = 2$, $\Delta d2_t = 1$ and $\Delta d3_t = 0$; for $t = 3$, $\Delta d2_t = -1$ and $\Delta d3_t = 1$. Re-writing equation (7) displays the intercept of the equation, which is a measure of the growth in performance of the control group:

$$\Delta Y_{it} = \alpha_0 + \alpha_3 d3_t + \gamma_1 D_G + \gamma_2 D_N + \gamma_3 \Delta X_{it3} + \dots + \gamma_k \Delta X_{itk} \Delta u_{it} \tag{8}$$

for $t = 2$ and $t = 3$, the estimates of the γ_j is identical in both equation (7) and (8).

Applying these growth models to those performance parameters which have passed the test of the linear model introduces a more rigourous test on the impact of FFS training.

Results

When considering the socioeconomic characteristics of farmers before participating in FFS, the non-FFS and the control groups in cropping year 1999/2000 (the year before they participated in the FFS training), their socioeconomic performances are similar in all three groups (Appendix 8.1).

Table 8.1 summarizes the results of the t-test comparing before and after differences for the three groups of farmers. More details of the results are shown in Appendix 8.2. For the FFS farmers, significant shifts were observed in all parameters except the gross margin of rice production. FFS farmers significantly reduced their pesticide use by 41.7 per cent, measured in quantity of active ingredient after the training, while no significant reduction was observed for the two other groups. Due to the pesticide reduction effect, the farmers' net benefit and the EIQ also showed significant differences. In addition to the quantity effect, the difference in the EIQ is influenced by a change in the type of pesticide used, that is, FFS farmers after the training opted for less toxic pesticides. Results for rice yields were less conclusive as they increased among all three groups of farmers. It must be recognized, however, that in intensive rice production systems, yield effects are difficult to attribute to IPM as several confounding factors such as adoption of new varieties can come into play. Therefore, no difference could be detected in gross margin where, in addition, changes in the use of other inputs can take place.

Two Period Growth Model

Based on the methodology outlined above, the analysis was proceeded by testing for change in performance in the growth rates of impact parameters. Here we included just two impact measures, namely quantity of pesticide use and EIQ. Gross margin were discarded from the econometric analysis because t-test results were non significant. Likewise, we did not include yield because of the somewhat ambiguous

TABLE 8.1
Summary of Short-term Linear Shift Effects from FFS Training

Farmer Group	Knowledge in Rice and Pest Management [score]	Yield [kg/ha]	Pesticide Use Reduction (gram a.i./ha) [$/ha]	Gross-margin [$/ha]	Farmer-net [$/ha]	Environmental Impact [score]
FFS	**	*	***	n.s.	**	***
Exposed	n.s.	*	n.s.	n.s.	n.s.	n.s.
Control	n.s.	**	n.s.	n.s.	n.s.	n.s.

Note: *, **, *** indicate the difference of before and after training at 0.10, 0.05 and 0.01.

t-test results. We did not include, either, results for farmer net benefits because this parameter is a combination of observed variables and an assumption borrowed from the literature. Of course, the results for this parameter will change as we change the assumption relating pesticide costs to health costs (Appendix 8.3).

The result of the two period growth model using the change in pesticide expenditures as the dependent variable show that FFS training has a significant effect on reducing farmers' pesticide use (see Table 8.2). This result is supported by the significant coefficient for rice and pest management knowledge. The positive sign of the constant term indicates that farmers' pesticide use is likely to continue to increase without FFS. Since the dummy variable for non-FFS is non-significant, there is no change in the trend of pesticide use among farmers living in the FFS village but not participating in the training. Summarizing the hypotheses tests in the lower panel of the table shows that a change in the positive trend in pesticide use is attributable to FFS. FFS farmers have significantly lower pesticide expenditures when compared to the non-FFS and control farmers on short-term (Table 8.2).

Using the environmental impact quotient as a dependent variable in the two period model also confirms the results of the t-test. FFS participation reduces the trend in the negative consequences of pesticides on the environment in the short-term (Table 8.2). As measured through the FFS participation dummy, the growth rate in EIQ of the FFS farmers shows a significant decline. It is also interesting to note that the counterfactual scenario (no FFS training) shows growing negative

TABLE 8.2
**Impact of FFS on Pesticide Expenditures and Environmental Impact
Quotient in the Short-term, Two Periods Growth Model**

Two Periods Growth Model	Δ in Pesticide Costs	Δ in EIQ
Constant (α)	0.248	2.340
	(1.576)	(3.096)***
Dummy for FFS (μ)	–0.485	–1.685
	(–2.368)**	(–1.715)*
Dummy for Non-FFS (β)	–0.220	–1.008
	(–0.937)	(–0.895)
Knowledge in rice and pest managements ($\Delta \ln K$)	–0.030	–0.133
	(–2.593)**	(–2.421)**
Total labour use ($\Delta \ln L$)	0.052	0.160
	(3.911)***	(2.498)**
R^2	0.109	0.064
F-statistics	7.236***	4.005***
Durbin-Watson statistic	1.853	1.883
N	241	241

Note: Data in parenthesis indicates the t-value. Pesticide expenditures are converted
to real value.

TABLE 8.3
**Impact of FFS on Pesticide Expenditures and Environmental Impact Quotient in
the Long-term, Three Periods Panel Data Growth Model**

Three Periods Panel Data Growth Model	Δ in Pesticide Costs	Δ in EIQ
Period 2 Dummy	–0.001	1.365
	(–0.006)	(2.798)***
Period 3 Dummy	0.077	–1.894
	(0.730)	(–4.063)***
Dummy for FFS	–0.254	–1.869
	(–2.167)**	(–3.616)***
Dummy for Non-FFS	0.137	0.041
	(1.219)	(0.068)
Knowledge in rice and pest management ($\Delta \ln K$)	–0.229	–0.073
	(–2.517)**	(–0.181)
Total labour use (man-day) ($\Delta \ln L$)	0.445	1.060
	(10.561)***	(5.698)***
R^2	0.448	0.294
F-statistics	28.183***	11.505***
Durbin-Watson statistic	1.817	1.467
N	188	188

Note: Data in parenthesis indicates the t-value. Pesticide expenditures are converted
to real value.

environmental impact from pesticides. This can be concluded from the intercepts of the models, which were significant at the 0.01 per cent level in the short-term. Again, within-village diffusion towards more environmentally benign pesticide use practices does not seem to be sustained as shown by the non-significant variable for non-FFS.

Three Period Growth Model

To test for the long-term effects of FFS training a three-period growth model (see Wooldridge, 2000) was used. Two time period dummy variables are included.

The long-term effects of FFS on farmer's pesticide use confirm the results of the short-term effect. Hence, FFS farmers retain their improved and more judicious pesticide use practices and continue to reduce pesticide use over time. By contrast, no significant change can be observed for the non-FFS farmers and the farmers in control villages in either period. Again change for both short and long-term knowledge had a significant effect on pesticide reduction.

For the EIQ variable the long-term change followed the results of pesticide use expenditures. On the long-term, FFS farmers not only reduce pesticide-use levels but also continue to adopt safer products. Knowledge seems to be a major driver for this process. On the other hand, no significant change can be observed for non-FFS farmers. The counterfactual scenario however shows ambiguous results as the first-time dummy variable shows a significant positive result, which in the second time period the trend is reversing. It is possible that these farmers may finally learn by experience the harmful effects from highly toxic pesticides and thus change their practices. Unfortunately, no information was collected in this direction.

Conclusions

Results of this study show that farmers who participated in the Farmer Field School training retain their knowledge and continue to practice improved IPM practices. Growth rates of pesticide expenditures and environmental impact are significantly reduced by the FFS training, both in the short- and long-term. On the other hand, farmers not trained in FFS tend to continue non-judicious ways of using chemical pesticides. Thus, the Farmer Field School approach is an effective method to reduce uneconomical use of chemical pesticides for rice production in Thailand. It can help farmers to sustainably change pesticide-use practices. However, the direct economic benefits of farmers expressed in terms of increased yields and gross margins could not be shown in this study. In high productivity rice production, such effects are difficult to detect and may indeed be small. Also, in rice, pesticide use does not account for a high share of the variable costs and therefore gross margin differences can be confounded by other factors.

However, it seems that when it comes to pesticide use, small farmers in developing countries may adopt new pest control methods even though the effects on profit may be low, if there are other benefits. Several studies have shown that small farmers show willingness to pay for safer pesticides (Cuyno et al., 2001; Garming and Waibel, 2006). In addition, it is also important to point out that changing farmer's pesticide-use practices generates additional environmental benefits that accrue to society at large. Finally, in the longer term, more judicious and better informed farmer crop and pest-management decision making can reduce the probability of pest outbreaks, These, however, did not occur during the years that the surveys were conducted.

Over all, the study suggests that a comprehensive assessment of the benefits of IPM that goes beyond only profit effects is necessary.

APPENDIX 8.1
Socioeconomic Characteristics of Sampling Farmers
by Group, before the FFS Training, Cropping Year 1999/2000

Socioeconomic Parameters	FFS before Training (n = 107)	Non-FFS (n = 58)	Control (n = 76)
No. of Household Members (persons)	4.95	4.86	4.39
No. of full time farm members (persons)	2.57	2.74	2.26
Age of Household Head (years)	48.31	49.90	47.14
Education (%)			
– primary school and lower	93.46	96.56	93.42
– high school	5.61	3.44	6.58
– greater than high school	0.93	0.00	0.00

(Appendix 8.1 Continued)

(*Appendix 8.1 Continued*)

Socioeconomic Parameters	FFS before Training (n = 107)	Non-FFS (n = 58)	Control (n = 76)
Land tenure (%)			
– owner	88.79	87.93	73.68
– rent	11.21	12.07	26.32
Household acquired credits (%)	63.55	62.07	67.11
Source of irrigation water (%)			
– irrigation canal	60.57	71.43	47.24
– creek	24.37	19.25	30.67
– ground water	3.59	1.24	11.04
– multiple sources	11.47	8.08	11.05
Off-farm income ($/household)	779.31	696.14	518.91
Farm size (ha)	2.89	3.72	2.73
Average number of plots	2.68	2.88	2.28
Rice area in % of farm area	84.25	87.23	85.75
% Number of rice crops per year			
– 1 crop	22.43	13.79	27.63
– 2 crops	43.93	43.10	38.16
– 3 crops	33.64	43.10	34.21
Rice yield (kg/ha)	3,496.48	3,529.73	3,181.56
Land preparation method (%)			
– tillage	67.29	60.34	68.42
– low tillage	32.71	39.66	31.58
Use of rice varieties (%)			
– modern variety	75.99	77.70	74.63
– local variety	24.01	22.30	25.37
Seed utilization (%)			
– own /neighbour	77.57	81.03	77.63
– purchase	22.43	18.97	22.37
Planting method (%)			
– transplanting	45.79	32.76	44.74
– wet direct seedling	44.86	56.90	42.11
– dry direct seedling	4.67	3.45	13.16
– multiple methods	4.68	6.89	0.00
Crop monitoring practices by farmer (%)	33.95	39.81	36.87
Pesticide Use (gram a.i./ha)	1,350.65	1,492.46	1,313.07
Pesticide Costs ($/ha)	23.98	27.37	22.48

(*Appendix 8.1 Continued*)

(*Appendix 8.1 Continued*)

Socioeconomic Parameters	FFS before Training (n = 107)	Non-FFS (n = 58)	Control (n = 76)
Pesticide use by WHO class (%)			
– extreme and high toxic (Ia and Ib)	56.08	65.52	53.95
– moderate toxic (II)	14.02	13.79	7.89
– unidentified in WHO class	1.87	1.72	2.63
– not using	28.04	18.97	35.53
Insecticide use by WHO class (%)			
– extreme and high toxic (Ia and Ib)	33.65	34.49	28.95
– moderate toxic (II)	14.02	15.52	10.53
– low toxic (III)	0.00	1.72	2.63
– not using	52.34	48.28	57.89
Fertilizer Use (kg of N/ha)	72.81	83.82	76.41
Fertilizer Costs ($/ha)	43.55	41.24	44.93
Cost of rice production ($/kg)	0.105	0.109	0.137
Labour input (md/ha)	57.29	51.22	69.69
Percentage of hired labour (%)	20.79	14.25	16.41
Knowledge of rice management (Score)	8.97	8.05	8.29
Gross margin ($/ha)	174.63	184.46	124.47

APPENDIX 8.2
**The Average Differences in Selected Performance
Indicators for Short-term Impact**

	Knowledge in Rice & Pest Management (Score)	Yield (kg/ha)	Total Pesticide (g.a.i./ha)	Gross Margin ($/ha)	Farmer Net Benefit ($/ha)	EIQ (Score)
FFS	2.05	209.66	−566.70	19.19	27.04	−3,761.46
n = 107	$(1.675)^*$	$(1.857)^*$	$(−4.279)^{***}$	$(1.395)^{ns}$	$(1.962)^{**}$	$(−3.603)^{***}$
Non-FFS	0.21	246.05	−170.50	3.25	9.21	−1,321.38
n = 58	$(0.134)^{ns}$	$(1.827)^*$	$(−1.270)^{ns}$	$(0.187)^{ns}$	$(0.522)^{ns}$	$(−1.374)^{ns}$
Control	−2.17	306.27	−154.48	5.37	9.19	−1,053.07
n = 76	$(−1.327)^{ns}$	$(2.567)^{**}$	$(−0.973)^{ns}$	$(0.386)^{ns}$	$(0.635)^{ns}$	$(−1.014)^{ns}$

Note: The difference is taken from the performance indicators of after training in
2000/2001 minus before training in 1999/2000.
The number in parenthesis is the t-value.

APPENDIX 8.3
Scenario Testing for the Farmer Net Benefit for the Short-term Impact

	Average Difference in Farmer Net Benefit for Short-term Impact ($/ha)		
Farmer Group	Ratio of Pesticide to Health Costs = 1:1	Ratio of Pesticide to Health Costs = 1:1.5	Ratio of Pesticide to Health Costs = 1: 2
FFS	27.04	165.25	193.06
$n = 107$	$(1.962)^{**}$	$(1.636)^{*}$	$(1.866)^{*}$
Non-FFS	9.21	5.29	7.52
$n = 58$	$(0.522)^{ns}$	$(0.044)^{ns}$	$(0.061)^{ns}$
Control	9.19	17.67	22.31
$n = 76$	$(0.635)^{ns}$	$(0.176)^{ns}$	$(0.218)^{ns}$

Note: The difference is taken from the farmer net benefit of after training in 2000/2001 minus before training in 1999/2000.
The number in parenthesis is the t-value.

References

Anderson, D.R., D.J. Sweeney and T.A. Williams. 2002. *Statistics for Business and Economics*. New York, USA: South-Western College Publishing.

Cuyno, L.C.M., G.W. Norton and A. Rola. 2001. 'Economic Analysis of Environmental Benefits of Integrated Pest Management: A Philippine Case Study'. *Agricultural Economics*, 25: pp. 227–33.

Erickson, R. 2004. 'Review and Evaluation Technical Assistance No. 3383-PAK: Integrated Pest Management', Asian Development Bank, Philippines.

Feder, G., R. Murgai and J.B. Quizon. 2003. 'Sending Farmers Back to School: The Impact of Farmer Field Schools in Indonesia', *Review of Agricultural Economics*, 26(1): pp. 1–18.

———. 2004. 'The Acquisition and Diffusion of Knowledge: The Case of Pest Management Training in Farmer Field Schools, Indonesia'. *Journal of Agricultural Economics*, 55(2): pp. 221–43.

Garming, H. and H. Waibel. 2006. 'Willingness to Pay to Avoid Health Risks from Pesticides: A Case Study from Nicaragua'. Paper to be presented at the 46th Annual Conference of GEWISOLA, Germany.

Godtland E., E. Sadoulet, A. de Janvry, R. Murgai and O. Ortiz. 2004. 'The Impact of Farmer Field School on Knowledge and Productivity: A Study of Potato Farmers in the Peruvian Andes'. *Economic Development and Cultural Change*, 53(1): pp. 63–92.

Greene, W.H. 2000. *Econometric Analysis*, Fourth edition. New Jersey: Prentice-Hall International.

Kenmore, P. 1996. 'Integrated Pest Management in Rice', in G. Persley (ed.), *Biotechnology and Integrated Pest Management*, pp. 76–97. Wallingford: CAB International.

Kovach, J.C., P.J. Degnl and J. Tette. 1992. 'A Method to Measure the Environmental Impact of Pesticides'. *New York's Food and Life Sciences Bulletin No. 139*, Cornell University, Ithaca, USA. (Accessed from http://www.nysipm.cornell. edu/publications/EIQ_Value03.pdf on 27 April 2006).

Ooi, P., S. Praneetvatakul, H. Waibel and G. Walter-Echols. 2005. 'The Impact of the FAO-EU IPM Programme for Cotton in Asia'. *A Publication of the Pesticide Policy Project*, Special issue publication series, pp. 123.

Praneetvatakul, S. and H. Waibel. 2003. 'Farm-level Economic Analysis of Farmer Field Schools in Integrated Pest Management', Poster paper presented at International Association of Agricultural Economics conference, 16–22 August 2003, Durban, South Africa.

Quizon, J., G. Feder and R. Murgai. 2001. 'Fiscal Sustainability of Agricultural Extension: The Case of the Farmer Field School Approach'. *Journal on International Agricultural and Extension Education* 8 (spring 2001): pp. 13–24.

Rola, A. and P. Pingali. 1993. *Pesticides, Rice Productivity, and Farmers' Health: An Economic Assessment*. Los Baños (Philippines), IRRI, pp. 100.

Rola, A.C., S.B. Jamias and J.B. Quizon. 2002. 'Do Farmer Field School Graduates Retain and Share What They Learn?—An Investigation in Iloilo, Philippines'. *Journal of International Agricultural and Extension Education*, Spring 2002, 9(1): pp. 65–76.

Tripp, R., M. Wijeratne and V.H. Piyadasa. 2005. 'What Should We Expect from Farmer Field Schools? A Sri Lanka Case Study'. *World Development*, 33(10): pp. 1705–20.

Van den Berg, H. 2004: 'IPM Farmer Field Schools: A Synthesis of 25 Impact Evaluations', Wageningen University, Report prepared for the FAO Global IPM Facility, Rome.

Wooldridge, J.M. 2000. *Introductory Econometrics*. Australia: South-Western College Publishing.

Yang, P., M. Iles, S. Yan and F. Jolliffe. 2005. 'Farmers' Knowledge, Perceptions and Practices in Ransgenic Bt Cotton in Small Producer Systems in Northern China'. *Crop Protection*, 24(3): 229–39.

About the Editors and Contributors

The Editors

Keijiro Otsuka

Keijiro Otsuka is currently Professor and Director of the FASID-GRIPS Joint Graduate Program jointly organized by the Foundation for Advanced Studies on International Development and the National Graduate Institute for Policy Studies in Tokyo. He holds an M.A. in Social Sciences from the Tokyo Metropolitan University and a Ph.D. from the University of Chicago. He is also the Chairman of the Board of Trustees of the International Rice Research Institute (IRRI) in the Philippines. He is the President-elect for 2006–09 of the International Association of Agricultural Economists. His area of research interest includes development economics, industrial development and environmental economics. He has received several prestigious awards for his outstanding research publications.

Kaliappa Kalirajan

Kaliappa Kalirajan is currently a Professorial Fellow at the Foundation for Advanced Studies on International Development and also Professor at the National Graduate Institute for Policy Studies in Tokyo. He holds an M.Sc. in Mathematical Economics, an M.Litt. in Econometrics from the Madurai University and a Ph.D. in Economics from the Australian National University. His area of research interest includes Sources of

Productivity Analysis, Regional Economic Groupings and International Trade and Macroeconometric Modeling. His work on the stochastic frontier production function has given him international recognition. He has several books and research papers in reputed journals to his credit.

The Contributors

Md Abdus Samad Azad

Md Abdus Samad Azad is currenlty doing his doctoral studies in the Department of Agricultural and Resource Economics, Faculty of Agriculture, Food and Resources at the University of Sydney, Australia. He has earlier worked with the Bangladesh Rice Research Institute in Bangladesh.

Mahabub Hossain

Mahabub Hossain is Head of the Social Sciences Division and Leader of Fragile Ecosystems Program at IRRI. Before joining IRRI in 1992, he was the Director General of the Bangladesh Institute of Development Studies, Dhaka. He received an M.A. in Economics from Dhaka University and a Ph.D. in Economics from Cambridge University, England. His major area of research interest is Rural Development Policies. He has been conducting research on understanding rural livelihoods and factors contributing to changes in income distribution and poverty focusing on Bangladesh, Eastern India, Vietnam and Thailand. He has been heavily involved in developing IRRI's Strategy and Medium Term Plans. He has authored nine books and over 100 papers in refereed journals and edited books.

G.S. Ananth

G.S. Ananth is currently Professor and Head of the Department of Agricultural Economics at the University of Agricultural Sciences in Bangalore, India.

P.G. Chengappa

P.G. Chengappa is currently the Vice-Chancellor of University of Agricultural Sciences, GKVK, Bangalore, and a Consultant for International Food Policy Research Institute at New Delhi. He received his Masters degree in Agriculture from the University of Agricultural Sciences, and holds a Ph.D. in Agricultural Economics from Indian Agricultural Research Institute. He has specialized in the research areas of agricultural productivity, agricultural marketing and cooperation, and agricultural extension services. He has several publications including research reports to various national and international organizations to his credit.

Aldas Janaiah

Aldas Janaiah is currently Professor (Agri. Econ.) & Head of School of Agri business Management (SABM) at the Acharya N.G. Ranga Agricultural University (ANGRAU) in Hyderabad, and Senior Scientist at the National Centre for Agricultural Economics and Policy Research in New Delhi. He holds an M.Sc. in Agricultural Economics from the Andhra Pradesh Agricultural University, which is now called Acharya N.G. Ranga Agricultural University, and a Ph.D. in Agricultural Economics from the Institute of Agricultural Sciences, Banaras Hindu University, Varanasi (India). He has served as International Team Leader for the World Bank-Asian Development Bank Joint Project 'Assessing the Poverty Impacts of Public Expenditures in Irrigation in Viet Nam', from November 2002–May 2004. His area of research interests are: the Livelihood Impacts of Agricultural Modernization, farm-level impacts of Transgenic Crop Varieties/GMOs and institutional Eeconomics of Agricultural R&D. He has published extensively in these areas in reputed national and international journals.

N.D. Macleod

N.D. MacLeod is currently a Principal Research Scientist at the Council of Scientific and Industrial Research Organization (CSIRO) Sustainable Ecosystems (Brisbane) in Australia. He holds an M.A. in Economics from the Simon Fraser University, Vancouver in Canada. His areas of specialization and expertise are: rangelands economics, landscape ecology, natural resource management and smallholder crop-livestock farming systems. He has undertaken several projects nationally and internationally in his field of expertise. He has several publications in reputed scientific journals to his credit.

S. Wen

S. Wen is currently Associate Professor and Deputy Director at the Red Soil Experimental Station in the Chinese Academy of Agricultural Sciences. He holds a Diploma in Science in Agriculture, Soil Science from the University of New England in Australia and is continuing his Ph.D. in Plant nutrition in Hunan Agricultural University. His area of specialization and expertise include Pasture Agronomy and Soil Science.

M. Hu

M. Hu is currenlty Professor in the College of Land Resources and Environment at the Jiangxi Agricultural University. She holds a Bachelor of Agriculture in Animal Husbandry from the Jiangxi University in China. Her research fields of specialization and expertise are Animal Husbandry Economics and Tourism Economics.

Russell M. Wise

Russell M. Wise is currently a senior scientist at the Council for Scientific and Industrial Research (CSIR), South Africa. He has a B.Sc. (Hons) in Environmental & Geographical Science (UCT, South Africa), a Post-Graduate Diploma in Natural Resource Economics and a Ph.D. in Agricultural and Natural Resource Economics (UNE, Australia). Russell has specialist skills in the development and application of bio-economic simulation models as decision-support tools in environmental management. He has experience in the development and application of such models in areas as diverse as The Control of Invasive Alien Species, The Development of Climate-Change Adaptation Strategies and Investigating the Economics of Carbon-sequestration in Land-use Systems.

Oscar J. Cacho

Oscar Cacho is Associate Professor of Agricultural and Resource Economics at the University of New England, where he teaches Bio-economics, Operations Research and Farm Management. He holds a B.Sc. on Marine Biology (UAM, Mexico), an M.Sc. in Fisheries and a Ph.D. in Economics (Auburn University, USA). In recent years, his research work has focused on topics such as Global Warming, Economics of Grazing Systems, Management of Invasive Species and Land Degradation.

Kei Kajisa

Kei Kajisa is currently an Agricultural Economist in the Social Science Division of the International Rice Research Institute (IRRI) in the Philippines. He concurrently holds Associate Professorship in the FASID-GRIPS Joint Graduate Program jointly organized by the Foundation for Advanced Studies on International Development and the National Graduate Institute for Policy Studies in Tokyo. He has an M.A. from the Graduate School of International Politics, Economics and Business, Aoyama Gakuin University and a Ph.D from the Department of Agricultural Economics, Michigan State University. His research interests include The Role of Social Capital in Economic Development, Analysis of Institutional Arrangements in Agrarian Markets and Sustainable Development Through Natural Resource Management by Local Communities. He has several publications in his area of expertise.

Jonna P. Estudillo

Jonna Estudillo is currently a Faculty Fellow at the Foundation for Advanced Studies on International Development and also Associate Professor at the National Graduate Institute for Policy Studies in Tokyo. She holds an M.A. in Economics from the University of the Philippines (Diliman) and a Ph.D. in Economics from the University of Hawaii. Her research interests include Schooling Progression of Children, Gender and Intra-household Allocation of Resources, Gender and Intergenerational Transfers of Wealth in Rural Communities, and Green Revolution in the Philippines focusing on its impact on productivity and income distribution. She has several publications in reputed journals and has co-authored books in her area of expertise.

Yasuyuki Sawada

Yasuyuki Sawada is Associate Professor in the Faculty of Economics in the Graduate School of Economics at the University of Tokyo in Tokyo, Japan. His research fields are Applied Microeconometrics, Development Economics and International Economics. He has published a number of research papers in these areas in internationally reputed academic and policy oriented journals. He has been a consultant to several national and international organisations such as World Bank and the Asian Developement Bank.

Suwanna Praneetvatakul

Suwanna Praneetvatakul is currently Assistant Professor at the Department of Agricultural and Resource Economics, Faculty of Economics, Kasetsart University, Bangkok. She received her Ph.D. in Agricultural Economics from the University of Hohenheim, Stuttgart, Germany. Her research interests include impact assessement of agricultural research, economics for sustainable agriculture and development and economic valuation of agricultural natural resources and environment. She has several publications in her area of research interests.

Hermann Waibel

Hermann Waibel is currently Professor of Agricultural and Development Economics (Chair: Agriculture, Environment and Development since 2001) at the Faculty of Economics and Business Administration, University of Hannover, Germany. He holds an M.Sc. in Agricultural Economics (Diplom-Agrarökonom) from the University of Hohenheim, Germany, and a Ph.D. in Agricultural Economics, University of Hohenheim, Germany. His area of research priorities are: development economics, impact assessment of international agricultural research, resource and environmental economics of horticultural and agricultural systems in developing countries with emphasis on Asia, economics of integrated pest management with emphasis on Southeast Asia, economics of organic agriculture/horticulture, economics of peri-/urban agriculture. He has several publications in reputed journals to his credit.

Index

ACIAR, 24, 62–64, 69, 76–80, 98–99
    project on economic assessment of forage-based cattle feeding, and its benefits to small-households, 63
    role in integrating outputs of research on cattle feeding and handling of forages, 62
agricultural
    growth performance evaluation practices, 20–22
        Solow 'residual' approach, shortcomings of, 20
    output growth, link with technology, 19
    research and development, investments in, current status, 46
    research institutes
        effort in transferring location-specific technology innovations to farmers, 18
    research systems, current status, 46
    sustainable technologies:
        combining scientific and indigenous knowledge, 19–26
    technology development, impact of research investment on, 25
agricultural technological innovations
    introduction, current approach, 17
    yield gaps:
        'constraints projects' to study, 18
        between farmers' yields and scientists' experiments, 18
agriculture
    role in alleviating rural poverty, 17
    share in GDP, 17

total output growth:
    decomposition into input growth, technical progress and technical efficiency improvement, 20, 21
    workforce, proportion dependent upon, 17
agroforestry in developing countries, 82
agroforests, 85
    capturing benefits offered by, 85
    definition of, 85
Aman rice crop in Bangladesh, 28–32, 28–43
    cropping pattern, 28
    cultivation under Bolon System, 29 (Figure)
    double transplantation system of crop establishment, 28
Ananth, G.S., 24
Australian Centre for International Agricultural Research. See ACIAR
Azad, Abdus, 24

Bechan bari, 29
beef production in China
    agronomic projects sponsored by ACIAR and Chinese Government, 62
    based on forages:
        regions identified for encouragement, 62
    during 2000, 61
    forage based:
        assessment of impact on smallholder farms, 67–68

conclusions, based on findings of scenarios tested, 78–79

dependencies between households and agricultural crafts/industrial activities, 64

discussion on economic profitability of each scenario tested, 76–78

economic model, 64–67

economic model, structure of, 65 (Figure)

feed year plan for small holder cattle production, 69 (Table)

feed year planning, 69

household surveys and target group, 63–64

Input Data module, 65

interdependencies between farm activities, 64

methodology, 63–67

Profit Measures module, 66

profit measures used, 66–67

Resource Reconciliation module, 65

results, for each scenario tested, 71–76

scenario models, 69–71

selected household parameter values for scenario modelling, 71 (Table)

forage production and cattle feeding strategies:

assessment of impact on smallholder farms, 63–79

growth rate during 2000, 61

public policy, changes in recent years, 61

share in total output of meat during 2000, 61

target markets, 61

Bolon, 29–43

Bolon and *Naicha* Systems of rice cultivation

difference in yield and other parameters, 40 (Table)

economics of Aman rice cultivation under, 39–40

Technical efficiency estimates of rice farms, 42 (Table)

variation in agronomic parameters practised in cultivation of, 39 (Table)

Bolon bari, 29–30

Bolon plot, 29–31, 38–39

Bolon system, 29–31, 38–43

advantages of, 30

crop management aspects

comparison with *Naicha* system, 38

farmers' perceptions:

advantages, 42–43

disadvantages, 43

inputs use and costs of production

comparison with *Naicha* system, 38–39

of Aman rice cultivation:

arguments against, 31

arguments for, 31

methodology, 29–32

research to study economics of:

characteristics of sample farmers and sampling method, 36

conclusions and policy implications, 43–44

land type and adoption of the system, 36–38

methodology adopted, 32–36

results and discussion, 36–40

Socio-Economic Characteristics of Sample Farmers, 37 (Table)

technical efficiency of, 40–42

Cacho, Oscar, 26

carbon

credits based on sequestered carbon, claiming of, 82

payments, 26

sequestering, 26

by trees, 81

sinks, in developing countries, 82

cattle husbandry in China, current status, 61

Centre of Excellence (COE) project of the National Graduate Institute for Policy Studies, 19

CERs, 83–84, 93

CERs (Certified Emission Reduction), 83

Chengappa, P.G., 24

China, beef production. *See* beef production in China

Clean Development Mechanism (CDM), 82

climate change, mitigating by trees, 81

continuous-cropping systems, 81

crop production and grain stocks in India expansion of, factors for, 46

crop-cattle technologies, 19, 22
impact on economic conditions of households in China, 24

crops in Karnataka state, major
total factor productivity growth, 56–57 (Table)

crop-tree technologies, 19, 22, 26

Data Envelopment Analysis (DEA), 34

Dhan bari, 30

'double transplantation' system of crop establishment methodology, 28
profitability and economic efficiency of, 24

Estudillo, Jonna, 25, 113–114, 118–122, 124, 130, 135–136

farmer field school (FFS), 137, 138, 147, 150, 151
approach:
benefits of, 137
criticisms against, 137
environmental and economic impact on crop and pest management prac-tices of rice in Thailand, 26

impact analyses, current:
advanced methods with panel data and impact indicators, 138–40
conclusions, 146–50
effect on pesticide expenditures of farmers, 146–50
model adopted for analysis, 140–42
results of, 143–46
results using three period growth model, 146
results using two period growth model, 143–46
projects, impact analyses done in the past shortcomings of, 138

'farmers' own yield gaps', 18–23
caused by technical inefficiencies, 22
definition of, 18
emanating from their use of chosen technology, 19
influenced by non-price and organizational factors, 22
reduction by agricultural research, 22

Faustman model, 84

Green Revolution technology
impacts on income and schooling of children, 25
introduction in Philippines, 113

greenhouse gas mitigation activities, 82

Hayami, 20

Hossain, Mohabub, 24

Hu and MacLeod Wen, 24

Integrated Pest Management (IPM), 137–151

International Association of Agricultural Economists (IAAE) conference, 26th, 19

Janaiah, Aladas, 25

Kajisa, Kei, 25

Kyoto emission limitations, 82
Kyoto Protocol, 26, 82
    Article 12, Clean Development
        Mechanism (CDM), 82
    Article 3.3, Land Use Change and
        Forestry (LUCF), 82
    incentives provided to developing
        countries to invest in greenhouse
        gas mitigation activities, 82

land degradation in southeast Asia,
    causes for, 81
Land-use Change and Forestry, LUCF, 82
land-quality indicators, 83
land use(s)
    alternative methods, optimal com-
        bination of, 81
    building relatively poor soil-carbon
        content, results and effects:
        factors for decision making, 96
    decision path, optimal for manage-
        ment:
        factors for decision making,
            95–97
    planting crops vs. growing trees, dif-
        ferent scenarios and management
        options, analysis:
        calibration of model and param-
            eter values, 85–89
    discussion and conclusions, 95–97
    issues requiring further analysis,
        96–97
    optimal decision rules and optimal
        state transitions, 89–95
    optimal decision rules, 89–91
    optimal state paths, 91–93
    sensitivity results, 93–95
    using meta model with dynamic
        programming (DP) algorithm,
        83–97
    when fuel prices are high, 96
    soil-carbon content, relatively high:
        factors for decision making, 95
    soil-carbon content, relatively poor:
        factors for decision making,
        95–96, 95

meat production in China, target mar-
    kets, 61
modern irrigation systems
    impact on livelihood of farmers with
        and without access to wells, 25
modern agricultural technology
    impact of adoption of, 25
    results on rice production, 113
MV (Modern Varieties) of rice irrigation
    benefits of, 113–14
    benefits to farmer cultivators, 113–14
    benefits to landless agricultural
        workers, 113–14
    benefits to urban consumers, 114
    impact on household-schooling
        investment decisions, 114
    conclusions and findings of hy-
        pothesis testing, 134
    description of sample children,
        119–21
    description of sample households,
        117–21
    determinants of children's school-
        ing and their occupational
        choice, 121–30
    determinants of credit constraints
        in the study villages of
        Philippines, 132–33
    econometric models used for
        hypothesis testing, 116–17
    hypothesis 1: adoption increases
        farm income and school
        investments, 114
    hypothesis 2: more educated
        children are more likely to
        engage in non-farm labour,
        114
    hypothesis 3: income effects on
        schooling investment have
        increased due to the develo-
        pment of non farm labour
        market, 114
    hypothesis testing, 114
    panel data set used for hypothesis
        testing, 114–15

probability to engage in non-farm work, 121–25

schooling progression of children, 125–30

structural interpretation of nexus between farm income and enhanced child schooling, 130–33

theoretical approach in hypothesis testing, 115–16

Naicha, 29–43

National Graduate Institute for Policy Studies

Centre of Excellence (COE) project, 19

NPV, 83, 84

Praneetvatakul, Suwanna, 26

present value of net revenues (NPV), 83

Production Possibility Frontier (PPF) of a local economy, 81

Red Soils region of China

forage based beef production systems: assessment of impact on smallholder farms, 63–79

research investment

impact on agricultural technology development and total factor productivity of major field crops, 25

impact on development of technologies and rate of return research study on, 47

in major field crops in Karnataka: discipline-wise technology released, 51–52

impact of research investment on input saving, 52–55

impact of yield augmenting technologies on land, 52–55

impact on rate of return, 53–55

impact on technology development, 51

research study to assess, 47–59

major Crops by SAUs in Karnataka

discipline-wise number of technologies, other than crop varieties developed and released, 59 (Table)

number of varieties deceloped and released, 58 (Table)

number of technologies, other than varieties developed and released, 58 (Table)

rice, drought-tolerant varieties, 25

rice crop in Bangladesh

contribution to source of livelihood, 28

cropping pattern, 28

share in agricultural value added, 28

Sawada, Yasuyuki, 25

SCUAF (Soil Changes under Agriculture, Agroforestry and Forestry), 85

shifting-cultivation systems, 81

small-scale forestry in developing countries, 82

Stochastic Varying Coefficients Frontier Approach (SVFA), 34

Swarna rice crop variety in Eastern India, 43

tank irrigation systems

current status in South India, 100

current status in Tamil Nadu

effect of massive diffusion of private wells, 100–01

tank irrigation systems in Tamil Nadu

effect of massive diffusion of private wells, hypotheses testing:

binary comparisons, 103–05

conclusions and policy implications, 110–11

differential impact of decline in collective management of tanks, 105–07

hypotheses to be tested, 101

impact of decline in collective management of tanks on poverty, 102–07

impact on rice profit among well users, 107–10

measuring collective management of tanks, 102–03

on farmers not having access to private wells, 101

rice yield and income of non-well users, 101

study site and data collection methods, 102

whether over exploitation of ground water results in profit on rice among well users, 101

effect of massive diffusion of private wells, concerns raised, 101

Technical inefficiency (TE), 20

total factor productivity (TFP), 20, 25, 47–56

tree-based systems as viable alternatives to crops

unwillingness of landholders to consider, causes for, 81

environmental and social benefits provided by, 81

Waibel, Hermann, 26

water management systems

using pumps and wells (modern), 100

using tanks (traditional), 100

with increasing water scarcity: questions raised, 100

Wen, MacLeod, 24

Wise, Russell, 26